AF770294

ROLE OF MICROBES IN RESTORATION ECOLOGY

ROLE OF MICROBES IN RESTORATION ECOLOGY

Edited by:

Dr Prashant Pant, Dr Pratibha Pant
and Dr Prerna Pandey

Delve
PUBLISHING

www.delvepublishing.com

Role of Microbes in Restoration Ecology

Dr Prashant Pant, Dr Pratibha Pant and Dr Prerna Pandey

Delve Publishing
2010 Winston Park Drive,
2nd Floor
Oakville, ON L6H 5R7
Canada
www.delvepublishing.com
Tel: 001-289-291-7705
001-905-616-2116
Fax: 001-289-291-7601
Email: orders@arclereducation.com

© 2018 Delve Publishing

ISBN: 978-1-77361-195-2 (Hardcover)

This book contains information obtained from highly regarded resources. Reprinted material sources are indicated. Copyright for individual articles remains with the authors as indicated and published under Creative Commons License. A Wide variety of references are listed. Reasonable efforts have been made to publish reliable data and views articulated in the chapters are those of the individual contributors, and not necessarily those of the editors or publishers. Editors or publishers are not responsible for the accuracy of the information in the published chapters or consequences of their use. The publisher assumes no responsibility for any damage or grievance to the persons or property arising out of the use of any materials, instructions, methods or thoughts in the book. The editors and the publisher have at- tempted to trace the copyright holders of all material reproduced in this publication and apologize to copyright holders if permission has not been obtained. If any copyright holder has not been acknowledged, please write to us so we may rectify.

Notice: Registered trademark of products or corporate names are used only for explanation and identification without intent of infringement.

Delve Publishing publishes wide variety of books and eBooks. For more information about Delve Publishing and its products, visit our website at www.delvepublishing.com.

DECLARATION

Some content or chapters in this book are open access copyright free published research work, which is published under Creative Commons License and are indicated with the citation. We are thankful to the publishers and authors of the content and chapters as without them this book wouldn't have been possible.

ABOUT THE EDITORS

Dr. Prashant Pant, is an Assistant Professor of Botany in Dyal Singh College, University of Delhi, New Delhi, India. He has been teaching for more than 5 years now and his main areas have been Bioinformatics, Molecular Biology and Genomics at the college. His PhD was on understanding the Microbial Succession using a Metagenomic approach in a Barren mined-out site undergoing ecological restoration. For this, he deployed many molecular and computational Biology tool and techniques. He has several publications (articles and book chapters) in International Journals and International Publications. Currently, he is also Chief Editor of Bioinformatics Review and a co-founder/Executive Chief Editor of a Science Journal in Hindi Language "Swadeshi Vigyan" for the native language speaking communities. Both the Journals are a service to the scientific community as they are fully open access.

Dr. Pratibha Pant, is an Assistant Professor of Botany in Daulat Ram College, University of Delhi, New Delhi, India and has a teaching experience of more than 6 years. She is a plant virologist and biotechnologist by training and has done her PhD in Characterization of Begomoviral genomes from Papaya from Department of Botany, University of Delhi. During her PhD, she utilized a combination of molecular biology and Computational Biology tools to analyze begomoviral genomes and to her credit, a couple of novel begomoviruses have been isolated, sequenced and reported in public databases.

Dr. Prerna Pandey, a biotechnologist with several years of wet lab research experience, worked at International Center for Genetic Engineering and Biotechnology, New Delhi. Her field of research in PhD included isolation and molecular characterization of geminiviruses, genome sequencing, gene annotation, and gene silencing using the RNA interference technology. She has also worked at Transasia Biomedicals and Advance Enzyme Technologies as a scientist. Prerna has published papers in peer-reviewed journals, and has submitted a number of annotated Geminiviral genome sequences in the GenBank, including two novel ones. She has also completed her editing and proof-reading courses from Society for the promotion of Editors and Proof-readers, UK and now works as a freelance scientific editor and writer. When Prerna is not busy with her assignments, she loves to explore historical places.

TABLE OF CONTENTS

LIST OF CONTRIBUTORS

Jennifer Wood

[1]Jennifer Wood is post-doctoral fellow in Dr. Ashley Franks' Applied and Environmental Microbiology Laboratory at La Trobe University, Australia. She has recently completed her Doctoral research in Microbial Ecology at the same university. Her broad research interest is in understanding the ecology of soil-microbial communities and their interactions with macro-communities. She is interested in developing trait-based approaches to interpret and predict how environmental and anthropogenic changes impact microbial community ecology and function.

Dr. Eleonora Egidi

[2] Dr. Eleonora Egidi is a mycologist interested in different aspects of fungal research, including fungal ecology, biology and phylogeny. Eleonora has an international background and has been involved in collaborative projects for the identification of new fungal barcoding regions, study of fungal communities from extreme environments, and use of fungi in phytoremediation. Her current scientific interests focus on exploring the biodiversity of Australian fungal flora and their role as drives of plant diversity in natural ecosystems, as well as understanding how these systems respond to environmental changes.

Dr. Prashant Pant

[3] Dr. Prashant Pant, is an Assistant Professor of Botany in Dyal Singh College, University of Delhi, New Delhi, India. He has been teaching for more than 5 years now and his main areas have been Bioinformatics, Molecular Biology and Genomics at the college. His PhD was on understanding the Microbial Succession using a Metagenomic approach in a Barren mined-out site undergoing ecological restoration. For this, he deployed many molecular and computational Biology tool and techniques. He has several publications (articles and book chapters) in International Journals and International Publications. Currently, he is also Chief Editor of Bioinformatics Review and a co-founder/Executive Chief Editor of a Science Journal in Hindi Language "Swadeshi Vigyan" for the native language speaking communities. Both the Journals are a service to the scientific community as they are fully open access.

Dr. Pratibha Pant

[4]Dr. Pratibha Pant, is an Assistant Professor of Botany in Daulat Ram College, University of Delhi, New Delhi, India and has a teaching experience of more than 6 years. She is a plant virologist and biotechnologist by training and has done her PhD

in Characterization of Begomoviral genomes from Papaya from Department of Botany, University of Delhi. During her PhD, she utilized a combination of molecular biology and Computational Biology tools to analyze begomoviral genomes and to her credit, a couple of novel begomoviruses have been isolated, sequenced and reported in public databases.

Dr Prerna Pandey

[5]Dr Prerna Pandey, a biotechnologist with several years of wet lab research experience, worked at International Center for Genetic Engineering and Biotechnology, New Delhi. Her field of research in PhD included isolation and molecular characterization of geminiviruses, genome sequencing, gene annotation, and gene silencing using the RNA interference technology. She has also worked at Transasia Biomedicals and Advance Enzyme Technologies as a scientist. Prerna has published papers in peer-reviewed journals, and has submitted a number of annotated Geminiviral genome sequences in the GenBank, including two novel ones. She has also completed her editing and proof-reading courses from Society for the promotion of Editors and Proof-readers, UK and now works as a freelance scientific editor and writer. When Prerna is not busy with her assignments, she loves to explore historical places.

Patrick D. Schloss

Department of Plant Pathology, University of Wisconsin–Madison, Madison, Wisconsin, United States of America

Jo Handelsman

Department of Plant Pathology, University of Wisconsin–Madison, Madison, Wisconsin, United States of America

Lucas William Mendes

Cell and Molecular Biology Laboratory, Center for Nuclear Energy in Agriculture CENA, University of São Paulo USP, Piracicaba, SP, CEP 13400-970, Brazil

Siu Mui Tsai

Cell and Molecular Biology Laboratory, Center for Nuclear Energy in Agriculture CENA, University of São Paulo USP, Piracicaba, SP, CEP 13400-970, Brazil

Franck O. P. Stefani

Department of Biological Sciences, Centre sur la biodiversité, Université de Montréal, Montréal, Québec, Canada

Terrence H. Bell

Department of Biological Sciences, Centre sur la biodiversité, Université de Montréal, Montréal, Québec, Canada

Charlotte Marchand
Department of Biological Sciences, Centre sur la biodiversité, Université de Montréal, Montréal, Québec, Canada

Ivan E. de la Providencia
Department of Biological Sciences, Centre sur la biodiversité, Université de Montréal, Montréal, Québec, Canada

Abdel El Yassimi
Department of Biological Sciences, Centre sur la biodiversité, Université de Montréal, Montréal, Québec, Canada

Marc St-Arnaud
Department of Biological Sciences, Centre sur la biodiversité, Université de Montréal, Montréal, Québec, Canada

Mohamed Hijri
Department of Biological Sciences, Centre sur la biodiversité, Université de Montréal, Montréal, Québec, Canada

Fisseha Asmelash
Forest and Range Land Biodiversity Conservation Directorate, Ethiopian Biodiversity Institute, Addis Ababa, Ethiopia
Department of Plant Biology and Biodiversity Management, Addis Ababa University, Addis Ababa, Ethiopia

Tamrat Bekele
Department of Plant Biology and Biodiversity Management, Addis Ababa University, Addis Ababa, Ethiopia

Emiru Birhane
Department of Land Resources Management and Environmental Protection, Mekelle University, Mekelle, Ethiopia
Department of Ecology and Natural Resource Management, Norwegian University of Life Sciences, Ås, Norway

Ed-Haun Chang
Mackay Junior College of Medicine, Nursing and Management, Beitou, Taipei 112, Taiwan

Tsai-Huei Chen
Taiwan Forestry Research Institute, Taipei 10066, Taiwan

Guang-long Tian
Environmental Monitoring and Research Division, Monitoring and Research Department, Metropolitan Water Reclamation District of Greater Chicago, IL 60804, USA

Chun-Kai Hsu
Taiwan Forestry Research Institute, Taipei 10066, Taiwan

Chih-Yu Chiu
Biodiversity Research Center, Academia Sinica, Nangang, Taipei 11529, Taiwan

PREFACE

The editors are glad to present this book on a concurrent and succinct topic "Role of Microbes in Restoration Ecology". Successful eco-restoration of degraded lands can provide essential ecosystem services benefiting the site, environment, wild flora and fauna as well as the locals. Establishment of native and economically-useful plant community in environmentally degraded areas requires rejuvenation of soil health and resetting of feedback loops that connect plant communities with microbial communities. Bacteria, fungi and soil microfauna in the soil constitute complex, heterogeneous communities that shift with changes in soil factors due to eco-restoration practices, enabling establishment of sustainable plant communities. The books presents a panoramic view of cultivable and uncultivable soil microbial biodiversity and the linkages associated between the above and below ground communities in the first chapter. The second chapter presents a deep insight on role of microbes as driver of soil processes. The next chapter focusses on the role of microbes as amelioration agents for polluted sites with various examples and case studies. Chapter four talks about the connection between techniques applied in eco-restoration programs for introduction and establishment of forest communities and the role played by microbial communities for success of such programs. Chapter five of the book dissects some of the most recent finding and developments regarding the interaction between soil microbial diversity with invasive alien species and the forces shaping microbial communities and current ecological theories. The last chapter presents some of the case studies where microbes have been fundamental in ecologically restoring the site in a short period of time. We believe that with this distribution of relevant and connected topics into chapters, the structure of the book will present a continuum of information on the role played by microbes and their importance in ameliorating and sustaining any ecosystem whether terrestrial or aquatic.

We look forward to and will really appreciate feedback by our readers.

Editor

1
CHAPTER

A PANORAMIC VIEW ON CULTURE DEPENDENT AND CULTURE INDEPENDENT MICROBIAL DIVERSITY

Jennifer Wood[1], Eleonora Egidi[2], Prashant Pant[3], Pratibha Pant[4], Prerna Pandey[5]

[1]Jennifer Wood is post-doctoral fellow in Dr. Ashley Franks' Applied and Environmental Microbiology Laboratory at La Trobe University, Australia. She has recently completed her Doctoral research in Microbial Ecology at the same university. Her broad research interest is in understanding the ecology of soil-microbial communities and their interactions with macro-communities. She is interested in developing trait-based approaches to interpret and predict how environmental and anthropogenic changes impact microbial community ecology and function.

[2] Dr. Eleonora Egidi is a mycologist interested in different aspects of fungal research, including fungal ecology, biology and phylogeny. Eleonora has an international background and has been involved in collaborative projects for the identification of new fungal barcoding regions, study of fungal communities from extreme environments, and use of fungi in phytoremediation. Her current scientific interests focus on exploring the biodiversity of Australian fungal flora and their role as drives of plant diversity in natural ecosystems, as well as understanding how these systems respond to environmental changes.

[3] Dr. Prashant Pant, is an Assistant Professor of Botany in Dyal Singh College, University of Delhi, New Delhi, India. He has been teaching for more than 5 years now and his main areas have been Bioinformatics, Molecular Biology and Genomics at the college. His PhD was on understanding the Microbial Succession using a Metagenomic approach in a Barren mined-out site undergoing ecological restoration. For this, he deployed many

molecular and computational Biology tool and techniques. He has several publications (articles and book chapters) in International Journals and International Publications. Currently, he is also Chief Editor of Bioinformatics Review and a co-founder/Executive Chief Editor of a Science Journal in Hindi Language "Swadeshi Vigyan" for the native language speaking communities. Both the Journals are a service to the scientific community as they are fully open access.

[4]Dr. Pratibha Pant, is an Assistant Professor of Botany in Daulat Ram College, University of Delhi, New Delhi, India and has a teaching experience of more than 6 years. She is a plant virologist and biotechnologist by training and has done her PhD in Characterization of Begomoviral genomes from Papaya from Department of Botany, University of Delhi. During her PhD, she utilized a combination of molecular biology and Computational Biology tools to analyze begomoviral genomes and to her credit, a couple of novel begomoviruses have been isolated, sequenced and reported in public databases.

[5]Dr Prerna Pandey, a biotechnologist with several years of wet lab research experience, worked at International Center for Genetic Engineering and Biotechnology, New Delhi. Her field of research in PhD included isolation and molecular characterization of geminiviruses, genome sequencing, gene annotation, and gene silencing using the RNA interference technology. She has also worked at Transasia Biomedicals and Advance Enzyme Technologies as a scientist. Prerna has published papers in peer-reviewed journals, and has submitted a number of annotated Geminiviral genome sequences in the GenBank, including two novel ones. She has also completed her editing and proof-reading courses from Society for the promotion of Editors and Proof-readers, UK and now works as a freelance scientific editor and writer. When Prerna is not busy with her assignments, she loves to explore historical places.

INTRODUCTION

Biological diversity is the foundation for the maintenance of ecosystems and its services (Loreau et al.,. 2001). A large proportion of the biological diversity within terrestrial systems is hidden below-ground in soils where microorganisms account for majority of the species pool. It has been estimated that one gram of soil contains as many as 10^{10}–10^{11} bacteria (Torsvik et al., 1990; Horner-Devine et al.,. 2003), between 6,000 and 50,000 bacterial species (Curtis et al.,. 2002), and about 200 million fungal hyphae (Leake et al.,. 2004). Such is the soil as microbial habitat that it harbors almost uncountable diversity of microbes. Soil microorganisms are arranged in diverse and dynamic communities that control a number of important ecosystem processes, including nutrient acquisition, nitrogen cycling, carbon cycling and soil formation, as well as aboveground diversity patterns (Burns, 1982; Hayat et al.,. 2010,

Wall and Moore, 1999). According to Fierer, 2017, **Soil Microbiome** is described as a general term describing all microorganisms that can be found in soil, such Bacteria (Eubacteria and Archaebacteria), Viruses, Fungi, Protists and other microbial eukaryotes such as microfaunal groups, microarthropods, worms, nematodes etc., (Fierer, 2017). Soil microorganisms are capable to alter rates of plant nutrient supply and resource partitioning via processes of nutrient mineralization, which facilitate the breakdown of soluble and insoluble organic matter into inorganic plant available forms (Van Der Heijden et al.,. 2007). At the plant-soil interface, bacteria known as plant growth-promoting rhizobacteria (PGPR) colonize the plant root-zone of influence (known as the rhizosphere) enhancing plant growth through a wide variety of mechanisms, such as nitrogen fixation, organic matter mineralization, control of soil-borne pathogens, and root growth promotion (Pii et al.,. 2015). Symbiotic associations between plants and nitrogen (N)-fixing bacteria are also crucial to convert atmospheric N into ammonium-N (Sprent, 2001), while mycorrhizal fungi enhance plant productivity by supplying limiting nutrients such Nitrogen, Phosphorous, Copper, Iron and Zinc to the plant, in exchange for Carbon (Read & Perez-Moreno 2003).

Given crucial role of microbial communities in regulating many important terrestrial processes, in this chapter we focus on discussing the functional and phylogenetic diversity of soil microorganisms, introducing some of the main concepts that will be further covered in chapters 2 and 3. We will discuss the diversity and distribution of soil microorganisms, as well as forces driving their community structure and occurrence. The elucidation of the extent of microbial biodiversity is crucial to better understand the functional capabilities of microbial communities, as well as predict their resilience and responses to environmental or anthropological perturbations. Understanding the dynamics of soil microbial communities in response to perturbation is the first step in elucidating how to harness soil microorganisms in bioremediation and restoration processes. Additionally, by understanding the changes in soil microbial communities it may be possible to predict the effects of environmental variations on many ecosystems and improve monitoring and conservation efforts.

TECHNOLOGICAL ADVANCEMENTS IN MICROBIAL DIVERSITY RESEARCH

The diversity of soil microorganisms has captured the attention of scientists since the earliest microbiological studies. The road to visualizing the diversity of microorganisms contained within the soil began in the 17th century with the advent of the microscope. By the 19th century culturing techniques, pioneered by Robert Koch, enabled the isolation and study of many environmentally and medically relevant bacteria in the laboratory (Ritz, 2007). However, discrepancies between what could be seen down the microscope and what could be grown in culture suggested a greater diversity of microorganisms than what could be studied using culturing techniques alone. Indeed, less than 1% of soil communities are culturable using known culturing techniques and nutrient media (Ritz 2007). This inability to investigate the majority of the soil microorganisms led to soil communities being described as the 'microbial black box'.

Nearly 100 years later, strategies for studying entire soil microbial communities began to emerge: in the 1980s phospholipid fatty acid analysis (PLFA) was applied successfully to investigate soil microorganisms and, although hotly debated, provided a useful approach for exploring and assessing broad community changes in response to environmental variables (Frostegård and Bååth 1996, Zelles, 1999, Frostegård et al.,. 2011). In the same decade, the development of the polymerase chain reaction (PCR) revolutionized the study of soil microbial ecology (Hirsch et al., 2010). In particular, PCR-based analysis of small subunit ribosomal RNA (rRNA) sequences enabled the development of new techniques for assessing phylogenetic diversity and community structure.

The rRNA molecule is particularly useful for identification and phylogenetic studies because it is universally present in all microorganisms and possesses both highly conserved and variable regions with different evolution rates (Figure 1). The comparison of 16S variable regions with existing sequences in databases, made the identification of new bacterial species more robust, especially for isolates that do not fit any recognized biochemical profiles (Rajendhran and Gunasekaran 2011). Similarly, the molecular identification of Fungi has also come to rely heavily on the nuclear ribosomal genes (Seifert et al.,. 2009; Schoch et al.,. 2012). The three main components of the fungal ribosomal operon include the large subunit (LSU, referred to as the 26S or 28S region, and including the two variable sub-regions D1 and D2), a small subunit (SSU or 18S), and the Internal Transcribed Spacer Region (ITS), comprising two hyper-variable regions (ITS1, ITS2) and a conserved 5.8S gene (Figure 1B). Recently, the whole ITS region has been ratified and accepted as the universal DNA marker for the species-level identification within the fungal kingdom (Schoch et al.,. 2012). The comparison of rRNA gene sequence similarities is now used as the 'gold standard' for taxonomic identification for Bacteria, Archea and Fungi, a concept known as "DNA barcoding".

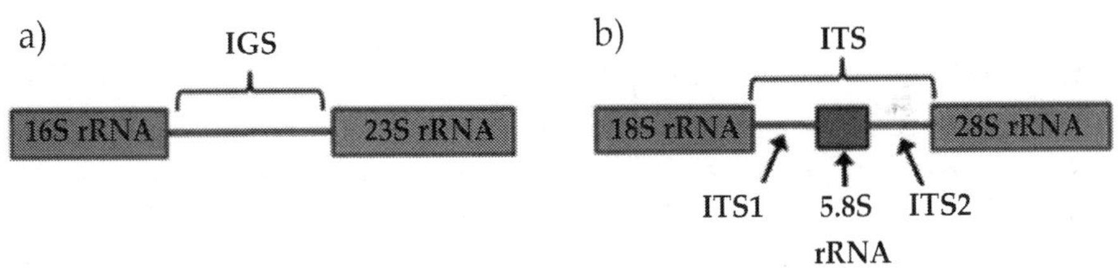

Figure 1: General schematic of rRNA genes used for molecular and phylogenetic analysis. (A) Bacterial rRNA operon: the variable non-coding intergenic spacer (IGS) region exhibits sequence and length polymorphisms that are exploited by community fingerprinting techniques such as ARISA, DGGE and TFRLP. The full length of the conserved 16S rRNA gene is used for phylogenetic analysis of bacterial species, whilst conserved and hypervariable domains within the 16S rRNA genes are utilized in next generation sequencing approaches. (B) Fungal rRNA operon: The fungal internal transcribes spacer (ITS) region exhibits sequence and length polymorphisms that are exploited by community fingerprinting techniques and its full length is used for taxonomic identification of fungal species.

After the pioneering work of Woese et al.,. (1987), PCR-based techniques using the rRNA operon to survey and rapidly visualise microbial community structure and diversity also started to be developed. Community fingerprinting techniques, such as denaturing/temperature gradient gel electrophoresis (DGGE/TGGE) (Muyzer, De Waal et al.,. 1993), terminal restriction fragment length polymorphism (TRFLP) (Liu, Marsh et al.,. 1997), and automated ribosomal intergenic spacer analysis (ARISA) (Fisher and Triplett 1999), were implemented to investigate complex or uneven communities. These approaches exploited sequence

(DGGE/TGGE, TFRLP) or length (ARISA) polymorphisms within the 16S intergenic spacer (IGS) and ITS regions of bacterial and fungal rRNA operons to visualize microbial community diversity in an entirely culture-independent manner (Figure 2a, b). Community fingerprinting allowed researchers to assess the effects of environmental variables on microbial communities via the comparison of the different community patterns across space or time. In parallel, the development of quantitative and real-time PCR techniques enabled the targeted enumeration of rRNA gene copy numbers within a community and, thus, microorganisms (Fierer, Jackson et al.,. 2005). Moreover, where known relationships between functional processes and genes existed, the quantification of community functional-potential or activity was possible (Denman and McSweeney 2006).

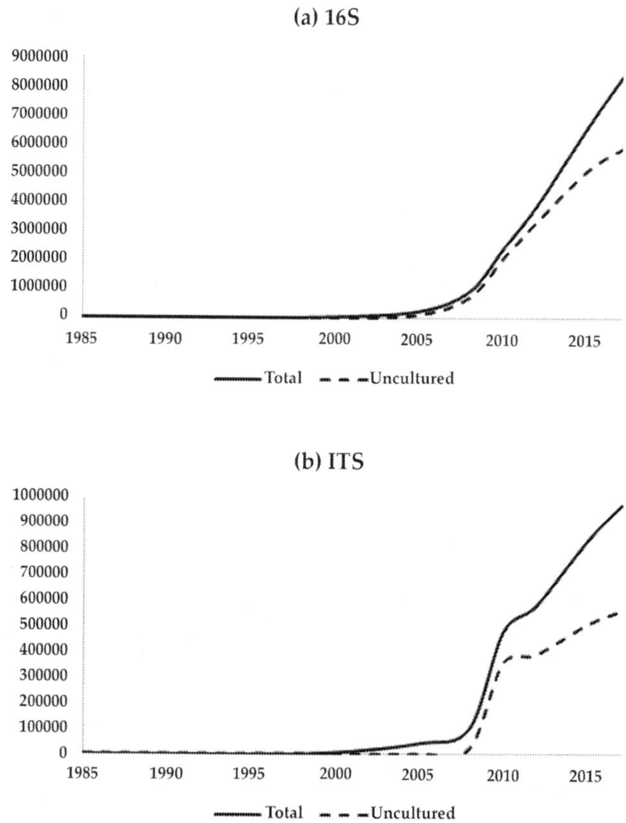

Figure 2: Graph showing the increase in number of (a) bacterial 16S and (b) fungal ITS sequences deposited in the INSDC database from 1985 to 2017.

The full and dashed lines indicate the number of total and uncultured organism sequences, respectively. Data accessed on 01.08.2017.

Thanks to the development of these culture-independent approaches, the responses of microorganisms to changing environments could be expanded to the not-cultivable fraction of the community, allowing for a more comprehensive understanding of the phylogenetic and functional diversity of microorganisms. However, although community fingerprinting and quantitative PCR techniques enabled ecologists to finally demonstrate both the effect of environmental variables on microbial communities and differences between communities, these methods were restricted in their ability to identify individual microbial species, limiting the studies to inferring community relationships. It is only with the advent of next-generation sequencing (NGS) in the 2000s that ecologists were able to truly look into the "microbial black box". Indeed, this novel technology has revolutionized the way microorganisms can be detected, studied, and discovered, offering the ability to produce an enormous volume (billions of short reads) of 16S or ITS data cheaply and in a single run (Schuster, 2008). Where previously community ecologists dealt in taxonomically ambiguous 'operational taxonomic units (OTUs)', names could be put to community members. Where community fingerprinting was limited to visualizing the most abundant community members, NGS enabled inspection of the rarest community members (Gobet et al., 2014, Van Dorst et al., 2014). Although many soil microorganisms remain unculturable, the unobservable diversity, identified a century before, is now observable, opening the way for novel opportunities to investigate soil microbes and their relationship with the environment.

ADVANCES IN UNDERSTANDING SOIL MICROBIAL PHYLOGENETIC DIVERSITY

In addition to allowing for the investigation of unculturable microbes, comparisons of rRNA genes also transformed the field of microbial taxonomy and systematics. Before the development of molecular techniques, many attempts were made to arrange or classify eukaryotic and prokaryotic microorganisms in distinct groups based on their morphological, biochemical and physiological similarities (Cohn 1875;

Migula 1897; Orla-Jensen 1909; Buchanan 1925; Kluyver and van Niel 1936; Chatton 1937; Stanier 1941; Stanier and van Niel 1962; Buchanan and Gibbons 1974; Murray 1986a, Seifert et al.,. 2009). However, the plasticity of many microscopic features, together with the morphological and physiological similarity among different microbial taxa, hampered the ability of these early attempts to establish and reliably classify the diversity of microbial species (van Niel 1946; Stanier and van Niel 1962; Stanier, Douroroff and Adelberg 1963).

Through the study of the molecular variability, evolutionary associations between different microbial species could be inferred in a systematic and reliable manner for the first time (Woese and Fox 1977; Fox et al.,.1980), with important repercussions for our understanding of microbial diversity. Remarkably, it is only thanks to the comparison of 16S bacterial rRNA hypervariable regions that the occurrence of two distinct Domains within the prokaryotes, the Bacteria and the Archaea, was recognised (Woese and Fox 1977; Woese 1987; Woese, Kandler and Wheelis 1990). This discovery radically shifted our view of evolution from a five kingdom to three domain (Bacteria, Archea, Eukaryotes) paradigm.

In the last three decades, a large amount of 16S and ITS sequence data has accumulated in public databases such as the INSDC (http://www.insdc.org/). At the time of writing this chapter, 8,318,788 (104,903 from type material) 16S and 975,956 (5,973 from type material) ITS sequences populate the INSDC repository (Figure 2).

Since its development in the mid 2000's, the massive use of next generation sequencing techniques to explore microbial community composition resulted in a steep increase in the amount of available sequences for both bacteria and fungi, with ~90% of the sequences having been generated in the last decade (Figure 2). This unprecedented amount of information based on the microbial genetic variability has greatly helped microbiologists reveal the true biodiversity of the "unseen majority". Indeed, recent estimates indicate that the sequencing efforts alone have allowed to recover about one-third of the bacterial and archaeal species (Schloss et al.,. 2016).

Remarkably, about half of all the bacterial and fungal sequences deposited in the INSDC database (4,782,373 and 553,581 for bacteria and fungi, respectively) are coming from "uncultured" organisms (Figure

2), many of which have not been previously isolated. A recent study by Schloss et al.,. (2016) reported that less than 20% of the microbial sequences have at least one cultured representative, while less than 1% of fungal species have been isolated with established protocols, indicating that many taxonomic groups are still undescribed and uncharacterized (Vartoukian et al.,. 2010; Tedersoo et al.,. 2016). Schloss et al.,. (2016) estimated that 26 out of the 65 recognized bacterial phyla lack a cultured representative. Similarly, Tedersoo et al.,. (2017) investigated the phylogenetic placement of previously unclassified soil fungi, revealing 37 novel major clades and seven single branches (singleton lineages) of uncultured taxa within the fungal kingdom. Interestingly, the distribution of many of these novel clades appear to be significantly influenced by climatic or edaphic conditions, suggesting that many undescribed fungal lineages with specific ecological preferences await discovery and formal description. Although studies based on single or multiple genetic markers have been crucial to reveal undescribed lineages of marine and terrestrial organisms among bacteria (Hugenholtz P et al.,. 1998) and fungi (Shadt et al.,. 2003), single-cell genomics and whole metagenome assays of the environment hold promise to be able to improve the phylogenetic placement and classification for many of these previously unknown organisms (Schoch et al.,. 2016, Jones et al.,. 2011, Green et al.,. 2005). These novel approaches enable the comprehensive sampling of all genes in all organisms, even in complex samples, and can yield insights into microbial diversity *and* the genetic potential of unculturable microorganisms which would be otherwise difficult or impossible to study. The most recent tree of life, constructed on the basis published sequences together with 1,011 newly reconstructed genomes that were recovered from a variety of environments, suggest that there exists as many as 92 bacterial and 26 archaeal phyla, many of which lack a formally described and cultured representative (Hag et al.,. 2016). With the aid of the metagenomic approach, the authors were able to identify a novel super-phylum, the Candidate Phyla Radiation (CPR), whose members are mainly symbionts and all share relatively small genomes and restricted metabolic capacities. This work highlights the importance of cultivation-independent genome-resolved strategies to access the large fraction of diversity carried by uncultivable microbial lineages, suggesting that there is still much biology to be uncovered in these unculturable groups.

CURRENT STATE OF KNOWLEDGE ON THE DIVERSITY AND DISTRIBUTION OF SOIL MICROORGANISMS

The application of rRNA sequencing has dispelled the myth that microorganisms isolated in pure culture from an environment represent the numerically dominant or functionally important species within a community. Rather, culturable microorganisms are, in fact, typically numerically rare within the communities from which they were obtained (Hugenholtz 2002). Prior to rRNA sequencing, culture based studies suggested that nine key bacterial genera were predominant in soils: *Agrobacterium*, *Alcaligens*, *Arthrobacter*, *Bacillus*, *Flavobacterium*, *Micromonospora*, *Nocardia*, *Pseudomonas* and *Streptomyces* (Alexander 1977, Janssen 2006). In contrast, 16S rRNA-based estimations of soil bacterial diversity indicated that collectively these nine genera represent < 4% of soil bacterial communities (Janssen 2006), with soil communities being typically dominated by alpha-Proteobacteria, beta-Proteobacteria, Actinobacteria and Acidobacteria (Table 1).

Global surveys of fungal diversity have revealed that Basidiomycota (42.3%) and Ascomycota (48.7%) represent the dominant fraction of many soil communities (Tedersoo *et al.,*. 2014). Within the Basidiomycota, members of the order Agaricomycetes are usually prevalent (38.7% of total fungal community), whilst within the Ascomycota, there are three dominant orders: Sordariomycetes (11.8%), Leotiomycetes (10.6%) and Eurotiomycetes (10.2%). Noticeably, the use of molecular markers has been instrumental in the recognition of a novel class of ubiquitous soil-inhabiting fungi, the Archaeorhizomycetes. Archaeorhizomycetes were firstly reported in tundra soils by Schadt et al.,. in 2003, where they were referred to as Soil Clone Group 1 (SCGI). However, the formal description of this group occurred only in 2011, when Rosling and colleagues were finally able to obtain the first isolate, *Archaeorhizomyces finlayi*. Previously, members of this clade were only known from ribosomal DNA sequencing, despite having been reported in more than 100 independent studies (Hibbett et al.,. 2016). This clade encompasses a wide variety of globally distributed soil and rhizospheric fungi whose ecological role has not been fully uncovered yet. Although it has been estimated than more than 145 species of Archaeorhizomycetes have been discovered using molecular surveys (Hibbett et al.,. 2016), only two have been formally described, *A. finlayi* and *A. borealis*, based on the only live cultures obtained so far. Globally, fungal communities are dominated by saprotrophs, which are important for the degradation of organic matter, followed by plant-associated ectomycorrhizal fungi (Tedersoo *et al.,*. 2014). Notably, for nearly 30% of the fungi, we still lack sufficient information for assigning functional groups (Table 2).

Table 1: Estimations of dominant bacterial phyla in soil using 21 16S rRNA clone librarians each with ≥90 clones (pre-NGS; (Janssen 2006) and NGS pyrosequencing of 9 grassland soils constituting 3 different land use types (post-NGS;(Will *et al.,*. 2010). Examples of each phylum/subphylum are also highlighted.

Phylum/Subphylum	Mean contribution (%)		Examples of genera
	Pre-NGS	Post-NGS	
α-Proteobacteria	19	11	*Sphingomonas, Rhizobium, Bradyrhizobium, Agrobacterium, Methylobacter, Nitrobacter, Nirospira, Rhodobacter*
β-Proteobacteria	10	18	*Burkholdaria, Nitrosospira, Thiobacillus*
γ-Proteobacteria	8	14	*Pseudomonas, Xanthamonas, Azotobacter, Thiocapsa, Chromatium*
δ-Proteobacteria	2	5	*Desulfovibrio, Bdellovibrio, Geobacter*
ε-Proteobacteria	<1	<0.25	*Helicobacter, Campylobacter*
Acidobacteria	20	19	*Acidobacterium*
Actinobacteria	13	13	*Arthrobacter, Rhodococcus, Streptomyces, Mycobacterium, Rubrobacter, Acidimicrobium*
Verrucomicrobia	7	1	*Chthoniobacter, Opitutus*
Bacteroidetes	5	2	*Chitinophaga, Sphingobacteria*
Firmicutes	2	4	*Clostridium, Bacillus, Lactobacillus*
Chloroflexi	3	1	

Planctomycetes	2	<0.25	
Gemmatimonadetes	2	1	*Gemmatimonas*
Unknown	2	9	

Table 2: Estimation of dominant fungal functional groups using pyrosequencing of soil collected from 365 sites across the globe (Tedersoo *et al.*,. 2014)

Functional group	Mean contribution (%)	Taxa associated with functional groups (common names in parentheses)
Saprotrophs	43.8	*Marasmius sp* (litter decomposers), *Armillaria spp.*(white rot), *Serpula lacrymans* (brown rot)
Ectomycorrhizal fungi	23.2	*Tuber* spp. (truffles), *Amanita* spp. (death caps)
Arbuscular Mycorrhizal fungi	0.5	phylum Glomeromycota
Mycoparasites	0.3	*Trichoderma spp., Stachybotrys elegans, Psathyrella epimyces*
Animal parasites	0.3	*Cordyceps spp., Entomopthora spp., Batrachochytrium dendrobatidis*
Plant pathogens	4	*Botrytis cinerea, Puccinia* spp. (rusts), *Fusarium graminearum, Fusarium oxysporum*
Other	27.8	

DRIVING FORCES OF SOIL MICROBIAL COMMUNITY STRUCTURE, DIVERSITY AND FUNCTION

The affordability of generating large-scale NGS datasets has enabled researchers to ask unique questions regarding the biogeography of soil microbial communities, as well as the processes governing their diversity, dispersal, and ecology (Fierer and Jackson 2006, Lauber *et al.,*. 2008, Tedersoo *et al.,*. 2010, Kivlin *et al.,*. 2011, Tedersoo *et al.,*. 2014). Additionally, large-scale, global collaborative initiatives, such as the TerraGenome project (Vogel *et al.,*. 2009) and the Earth Microbiome project (EMP) (Gilbert *et al.,*. 2014), which aim to construct a global catalogue of the uncultured microbial diversity of this planet, have significantly contributed to our understanding of soil microbial communities.

Such large-scale microbiological surveys have shed new light onto the relationships between micro- and macro-community diversity. Global spatial parameters (distance from equator) and climatic variables (mean annual precipitation) have been shown to be strong drivers of fungal species richness (Tedersoo *et al.,*. 2014), with dispersal limitations at both small and large scales being an important driver of fungal community assembly in soils (Peay et al.,. 2010). For bacteria these high-level micro-macro relationships are less clear. Whilst relationships between plants and bacteria clearly exist and, will be discussed in later chapters, continental-scale bacterial surveys indicate that there is no clear relationship between bacterial species richness and plant species richness. Nor is bacterial species richness related to variables that typically govern macro-community richness, such as temperature, and latitude (Fierer and Jackson 2006). Rather, soil pH proved to the be strongest predictor of bacterial species richness, with low-pH soils exhibiting the least diverse bacterial communities (Fierer and Jackson 2006). The observation that pH is a strong driver of bacterial community structure has been reported independently at various spatial scales and across multiple soil types with Acidobacteria subgroup 1 typically dominating communities in low pH soil and the alpha-Proteobacteria dominating in high pH soils (Lauber *et al.,*. 2009, Rousk *et al.,*. 2010). Interestingly, pH appears to have little impact on fungal community structure but does cause a turnover in growth activity between the Eukaryotic and Prokaryotic fractions of the community with bacterial growth decreasing and fungal growth increasing at lower pHs (Rousk *et al.,*. 2009). Interestingly, a recent meta-analysis based on 325 local community studies demonstrated that microclimatic variation within sites may hide the influence of macro-climatic variations among sites, suggesting that local scale processes may outcompete regional processes in determining microbial diversity and abundance patterns (Hendershot et al.,. 2017).

Although there is no clear relationship between plant-community diversity and bacterial-community diversity, both bacterial and fungal communities are strongly influenced by vegetation dynamics. For example, comparisons of temperate grasslands from four continents suggested that whilst microbial and plant community alpha diversity were not correlated, the levels of community turnover (beta-diversity) within plant, bacterial and fungal communities were significantly positively associated (Prober *et al.,*. 2015). This positive relationship was also observed via NGS comparisons of semi-arid plant patches and adjacent low-cover areas (Goberna *et al.,*. 2016), where plant richness and phylogenetic diversity were shown to have a positive impact on soil fertility, which in turn increased bacterial richness. Interestingly, increased soil fertility also decreased bacterial phylogenetic diversity suggesting that high levels of nutrient input by plants promote the dominance of lineages with high relative fitness under carbon-enriched condition. Recent studies have also demonstrated that soil biological diversity (prokaryotic and eukaryotic combined) is a strong predictor of soil fertility and plant productivity (as opposed to plant community diversity;(Delgado-Baquerizo *et al.,*. 2017), suggesting that preserving soil biodiversity at each level (from micro-to macro- scales) may positively impact nutrient cycling and plant diversity.

CONCLUSIONS

Microorganisms in soil are responsible for many fundamental processes at the ecosystem level and represent a crucial component of the soil biodiversity. Over the years, the use of universally present genes (or DNA barcodes) to identify microbial species has been instrumental in uncovering the true extent of this diversity, revealing several divergent but undescribed lineages within prokaryotes and eukaryotes, as well as numerous distinct Bacterial, Archeal and Fungal phyla among the unculturable fraction of soil microbial communities. Moreover, new technologies, such as single-cell genomes, are expected to further aid in revealing this diversity (Schloss et al.,. 2016, Tedersoo et al.,. 2017, Hag et al.,. 2016), while large-scale, global collaborative initiatives have helped reveal patterns of microbial diversity and community responses to many environmental variables. An increasing body of research is demonstrating that microbial and plant communities intimately connected, suggesting that microbial communities should be taken into account when designing environmental monitoring and remediation strategies.

SELECTED REFERENCES

1. Alexander, M. (1977). Introduction to soil microbiology, John Wiley & Sons.
2. Burns, R. G. (1982). Enzyme activity in soil: location and a possible role in microbial ecology. Soil Biology and Biochemistry 14(5), 423-427.
3. Delgado-Baquerizo, M., Powell, J. R., Hamonts, K., Reith, F., Mele, P., Brown, M. V., et al. (2017). Circular linkages between soil biodiversity, fertility and plant productivity are limited to topsoil at the continental scale. New Phytologist 215(3): 1186-1196.
4. Denman, S. E. and C. S. McSweeney (2006). Development of a real-time PCR assay for monitoring anaerobic fungal and cellulolytic bacterial populations within the rumen. FEMS Microbiology Ecology 58(3): 572-582.
5. Fierer, N. and Jackson, R. B. (2006). The diversity and biogeography of soil bacterial communities. Proceedings of the National Academy of Sciences of the United States of America 103(3): 626-631.
6. Fierer, N., J. A. Jackson, R. Vilgalys and R. B. Jackson (2005). Assessment of soil microbial community structure by use of taxonspecific quantitative PCR assays. Applied and Environmental Microbiology 71(7): 4117-4120.
7. Fisher, M. M. and E. W. Triplett (1999). Automated approach for ribosomal intergenic spacer analysis of microbial diversity and its application to freshwater bacterial communities. Applied and Environmental Microbiology 65(10): 4630-4636.

8. Frostegård, A. and E. Bååth (1996). The use of phospholipid fatty acid analysis to estimate bacterial and fungal biomass in soil. Biology and Fertility of Soils 22(1-2): 59-65.
9. Frostegård, Å., A. Tunlid and E. Bååth (2011). Use and misuse of PLFA measurements in soils. Soil Biology and Biochemistry 43(8): 1621-1625.
10. Gilbert, J. A., Jansson, J. K. and Knight, R. (2014). The Earth Microbiome project: Successes and aspirations. BMC Biology 12(1).
11. Goberna, M., Navarro-Cano, J. A. and Verdú, M. (2016). Opposing phylogenetic diversity gradients of plant and soil bacterial communities. Proceedings. Biological Sciences 283(1825): 20153003.
12. Gobet, A., A. Boetius and A. Ramette (2014). Ecological coherence of diversity patterns derived from classical fingerprinting and Next Generation Sequencing techniques. Environmental Microbiology 16(9): 2672-2681.
13. Green Tringe S., von Mering C., Kobayashi A., Salamov A. A., Chen K., Chang H. W., et al. (2005). Comparative metagenomics of microbial communities. Science 308:554–7
14. Hayat, R., Ali, S., Amara, U., Khalid, R., & Ahmed, I. (2010). Soil beneficial bacteria and their role in plant growth promotion: a review. Annals of Microbiology 60(4), 579-598.
15. Hendershot, J. N., Read, Q. D., Henning, J. A., Sanders, N. J. and Classen, A. T. (2017). Consistently inconsistent drivers of microbial diversity and abundance at macroecological scales. Ecology 98(7): 1757-1763.
16. Hirsch, P. R., T. H. Mauchline and I. M. Clark (2010). Culture independent molecular techniques for soil microbial ecology. Soil Biology and Biochemistry 42(6): 878-887.
17. Hugenholtz P., Goebel M., Pace, N. R. (1998). Impact of cultureindependent studies on the emerging phylogenetic view of bacterial diversity. Applied Environmental Microbiology. 180:4765–7
18. Hugenholtz, P. (2002). Exploring prokaryotic diversity in the genomic era. Genome Biology 3(2): 0003.0001-0003.0008.

19. Janssen, P. H. (2006). Identifying the dominant soil bacterial taxa in libraries of 16S rRNA and 16S rRNA genes. Applied and Environmental Microbiology 72(3): 1719-1728.
20. Jones MDM, Forn I, Gadelha C, Egan MJ, Bass D, Massana R, Richards TA. (2011). Discovery of novel intermediate forms redefines the fungal tree of life. Nature 474: 200–3.
21. Kivlin, S. N., Hawkes, C. V. and Treseder, K. K. (2011). Global diversity and distribution of arbuscular mycorrhizal fungi. Soil Biology and Biochemistry 43(11): 2294-2303.
22. Lauber, C. L., Hamady, M., Knight, R. and Fierer, N. (2009). Pyrosequencing-based assessment of soil pH as a predictor of soil bacterial community structure at the continental scale. Applied and Environmental Microbiology 75(15): 5111-5120.
23. Lauber, C. L., Strickland, M. S., Bradford, M. A. and Fierer, N. (2008). The influence of soil properties on the structure of bacterial and fungal communities across land-use types. Soil Biology and Biochemistry 40(9): 2407-2415.
24. Liu, W. T., T. L. Marsh, H. Cheng and L. J. Forney (1997). Characterization of microbial diversity by determining terminal restriction fragment length polymorphisms of genes encoding 16S rRNA. Applied and Environmental Microbiology 63(11): 45164522.
25. Loreau, M., Naeem, S., Inchausti, P., Bengtsson, J., Grime, J. P., Hector, A., et al. (2001). Biodiversity and ecosystem functioning: current knowledge and future challenges. Science 294(5543): 804-808.
26. Muyzer, G., E. C. De Waal and A. G. Uitterlinden (1993). Profiling of complex microbial populations by denaturing gradient gel electrophoresis analysis of polymerase chain reaction-amplified genes coding for 16S rRNA. Applied and Environmental Microbiology 59(3): 695-700.
27. Prober, S. M., Leff, J. W., Bates, S. T., Borer, E. T., Firn, J., Harpole, W. S., et al. (2015). Plant diversity predicts beta but not alpha diversity of soil microbes across grasslands worldwide. Ecology Letters 18(1): 85-95.
28. Rajendhran, J. and Gunasekaran, P. (2011). Microbial phylogeny and diversity: Small subunit ribosomal RNA sequence analysis and

beyond. Microbiological Research 166(2): 99-110.

29. Ritz, K. (2007). The plate debate: Cultivable communities have no utility in contemporary environmental microbial ecology. FEMS Microbiology Ecology 60(3): 358-362.

30. Rosling A, Cox, F., Cruz-Martinez, K., Ihrmark, K., Grelet, G, Lindahl B. D., et al. (2011). Archaeorhizomycetes: Unearthing an ancient class of ubiquitous soil fungi. Science 333 (6044): 876–9.

31. Rousk, J., Bååth, E., Brookes, P. C., Lauber, C. L., Lozupone, C., Caporaso, J. G., et al. (2010). Soil bacterial and fungal communities across a pH gradient in an arable soil. ISME Journal 4(10): 1340-1351.

32. Rousk, J., Brookes, P. C. and Bååth, E. (2009). Contrasting soil pH effects on fungal and bacterial growth suggest functional redundancy in carbon mineralization. Applied and Environmental Microbiology 75(6): 1589-1596.

33. Schadt, C. W., Martin, A. P., Lipson, D. A. (2003). Schmidt SK. Seasonal dynamics of previously unknown fungal lineages in tundra soils. Science 301:1359–61.

34. Schloss, P. D., Girard, R. A., Martin, T., Edwards, J., Thrash, J. C. (2016). mBio 7:e00201-16, 2016, http://dx.doi.org/10.1128/mBio.00201-16

35. Tedersoo, L., Bahram, M., Põlme, S., Kõljalg, U., Yorou, N. S., Wijesundera, R., et al. (2014). Fungal biogeography. Global diversity and geography of soil fungi. Science (New York, N.Y.) 346(6213): 1256688.

36. Tedersoo, L., May, T. W. and Smith, M. E. (2010). Ectomycorrhizal lifestyle in fungi: Global diversity, distribution, and evolution of phylogenetic lineages. Mycorrhiza 20(4): 217-263.

37. Van Dorst, J., A. Bissett, A. S. Palmer, M. Brown, I. Snape, J. S. Stark, B. Raymond, J. McKinlay, M. Ji, T. Winsley and B. C. Ferrari (2014). Community fingerprinting in a sequencing world. FEMS Microbiology Ecology 89(2): 316-330.

38. Vartoukian S., Palmer, R. M., Wade, W. G. (2010). “Strategies for culture of ‘unculturable’ bacteria”. FEMS Microbiol Ecology 309:1–7. 8.

39. Vogel, T. M., Simonet, P., Jansson, J. K., Hirsch, P. R., Tiedje, J. M., van Elsas, J. D., et al. (2009). TerraGenome: A consortium for the sequencing of a soil metagenome. Nature Reviews Microbiology 7(4): 252.
40. Wall, D. H., & Moore, J. C. (1999). Interactions underground: soil biodiversity, mutualism, and ecosystem processes. BioScience 49(2), 109-117.
41. Will, C., Thürmer, A., Wollherr, A., Nacke, H., Herold, N., Schrumpf, M., et al. (2010). Horizon-specific bacterial community composition of german grassland soils, as revealed by pyrosequencing-based analysis of 16S rRNA genes. Applied and Environmental Microbiology 76(20): 6751-6759.
42. Zelles, L. (1999). Fatty acid patterns of phospholipids and lipopolysaccharides in the characterisation of microbial communities in soil: a review. Biology and Fertility of Soils 29(2): 111-129.

2

CHAPTER

MICROORGANISMS AS DRIVERS OF SOIL ECOLOGICAL PROCESSES

Jennifer Wood[1], Eleonora Egidi[2], Prashant Pant[3], Prashant Pant

[1]Jennifer Wood is post-doctoral fellow in Dr. Ashley Franks' Applied and Environmental Microbiology Laboratory at La Trobe University, Australia. She has recently completed her Doctoral research in Microbial Ecology at the same university. Her broad research interest is in understanding the ecology of soil-microbial communities and their interactions with macro-communities. She is interested in developing trait-based approaches to interpret and predict how environmental and anthropogenic changes impact microbial community ecology and function.

[2] Dr. Eleonora Egidi is a mycologist interested in different aspects of fungal research, including fungal ecology, biology and phylogeny. Eleonora has an international background and has been involved in collaborative projects for the identification of new fungal barcoding regions, study of fungal communities from extreme environments, and use of fungi in phytoremediation. Her current scientific interests focus on exploring the biodiversity of Australian fungal flora and their role as drives of plant diversity in natural ecosystems, as well as understanding how these systems respond to environmental changes.

[3] Dr. Prashant Pant, is an Assistant Professor of Botany in Dyal Singh College, University of Delhi, New Delhi, India. He has been teaching for more than 5 years now and his main areas have been Bioinformatics, Molecular Biology and Genomics at the college. His PhD was on understanding the Microbial Succession using a Metagenomic approach in a Barren mined-out site undergoing ecological restoration. For this, he deployed many molecular and computational Biology tool and techniques. He has several publications

(articles and book chapters) in International Journals and International Publications. Currently, he is also Chief Editor of Bioinformatics Review and a co-founder/Executive Chief Editor of a Science Journal in Hindi Language "Swadeshi Vigyan" for the native language speaking communities. Both the Journals are a service to the scientific community as they are fully open access.

[4]Dr. Pratibha Pant, is an Assistant Professor of Botany in Daulat Ram College, University of Delhi, New Delhi, India and has a teaching experience of more than 6 years. She is a plant virologist and biotechnologist by training and has done her PhD in Characterization of Begomoviral genomes from Papaya from Department of Botany, University of Delhi. During her PhD, she utilized a combination of molecular biology and Computational Biology tools to analyze begomoviral genomes and to her credit, a couple of novel begomoviruses have been isolated, sequenced and reported in public databases.

INTRODUCTION

Soils are essential for life. Not only do they provide habitat for animal and insects and growing medium for plants, but they filter surface waters and store carbon (Brady, 2010). Soils are a mixture of living (biotic) and non-living (abiotic) components and consist of three main parts: (i) the mineral component, which constitutes about 45% of the soil by volume; (ii) the space between these particles which is made up of equal parts of air and water and collectively constitutes about 50% of the soil volume; and (iii) the organic matter, which comprises only about 5% of the soil (Brady, 2010). By volume, about 10% of the organic fraction is made up of plant roots, and just 5% is made of living organisms such as tiny insects, fungi and bacteria (Blume 2016). Despite contributing to less than 0.25% to the total soil volume, the estimated diversity of soil microorganisms is staggering. It is thought that 1 g of soil contains up to 10^{10}–10^{11} bacterial cells, 6000–50 000 bacterial species, and up to 200 m fungal hyphae (Van Der Heijden *et al.* 2008).

Soil microorganisms play a fundamental role in both below- and aboveground processes. Below-ground microorganisms contribute to the cycling of important nutrients including carbon (C), nitrogen (N) and phosphorus (P), which will be discussed in this chapter (Van Der Heijden *et al.* 2008). Soil microorganisms can also contribute to soil formation, soil structure, and the degradation of contaminants within the soil (discussed in chapter 5). Moreover, microbial interactions that take place at the plant-soil interface, also known as the rhizosphere, can govern aboveground plant community dynamics. Rhizosphere communities can harbor both plant-growth-promoting and disease-causing microorganisms, the delicate interplay between which can determine the fate of the plant (Raaijmakers *et al.* 2009). These plant-microbe interactions can help facilitate succession events and may even govern the composition of entire plant communities (Augspurger 1984, Van Der Heijden *et al.* 2008, Sonkoly *et al.* 2017). Therefore, understanding the microbial contribution to C cycling and plant nutrition represents an opportunity for manipulating specific microorganisms to enhance nutrients availability in soil.

MICROORGANISMS AND THE CARBON CYCLE

The cycling of carbon between soils and the atmosphere is a balance between photosynthesis and respiration (Brewer 1988). Photosynthesis fixes atmospheric carbon (CO_2), incorporating it into the organic matter and sugars that make up plant tissues. Organisms capable of fixing atmospheric CO_2 are called autotrophs. Photoautotrophs, such as plants, derive their energy from light and can synthesize all of their necessary organic compounds from CO_2. It is estimated that 5-21% of all the carbon fixed by plants is deposited directly into the soil in the form of root exudates (Lugtenberg and Kamilova 2009).

Although plants are primarily responsible for photosynthesis in terrestrial environment, soil microorganisms have a major role to play in the global carbon cycle. Estimates suggest that globally the amount of carbon stored in the organic component of soils is at least three times the amount of carbon stored in the atmosphere (Singh *et al.* 2010). Photoautotrophic and chemoautotrophic microorganisms (the latter of which derived their energy from chemical reactions rather than light) can contribute directly to the fixation of atmospheric carbon. Terrestrial microorganisms such as cyanobacteria have a major role as primary producers in some ecosystems, particularly on alkali soil (Dodds *et al.* 1995; Sonkoly *et al.* 2017). However, in an indirect way, microorganisms can impact atmospheric carbon fixation also, via their roles as plant symbionts or pathogens which can alter plant nutrient availability and plant respiration rates (Van Der Heijden *et al.* 2008).

In the soil, microorganisms transform and decompose organic matter - a process known as mineralization. Soil organic material includes root exudates and deceased plant and animal materials, and is decomposed by 'organic carbon-consuming' heterotrophic soil microorganisms (Brewer 1988). Heterotrophic soil microorganisms heavily contribute to the respiration of soil organic matter back into CO_2, but are also crucial to the degradation of complex dead organic matter into forms that can be reused by other organisms in the food web. Particularly important are wood-rotting fungi. Dead wood represents the major pool of carbon within the soil. Typically, the composition of wood is approximately 25% lignin, a structural molecule which can be consumed by very few organisms, making wood highly resistant to decay (Floudas *et al.* 2012; Novaes *et al.* 2010). The only organisms capable of substantial lignin decay are the wood-rot fungi from the class Agaricomycetes (Basidiomycota) (Floudas *et al.* 2012). The composition of the bacterial community can also affect carbon cycling as different bacterial phyla are associated with different rates of C mineralization: rapid growing (copiotrophic) beta-Proteobacteria and Bacteroidetes increase C mineralization whereas slow growing (oligotrophic) Acidobacteria is associated with slow C mineralization rates (Fierer *et al.* 2007).

In aerobic soils, the main microbial respiration product is CO_2. However, in anoxic soils, such as waterlogged peatlands or swamps and paddy fields, methane (CH_4) is formed from acetate, CO_2 and H_2 through the metabolic activity of methanogenic archaea (Thauer *et al.* 2008). It is estimated that methanogenic archaea produce approximately 1 billion tons of methane each year (Thauer *et al.* 2008). The carbon cycle does not operate independently but it is closely coupled to the cycling of other essential elements used in microbial metabolism. This linkage occurs either via the use of the other elements as electron donors and acceptors in energy transduction (e.g. nitrogenous and sulphur compounds).

MICROORGANISM AND THE NITROGEN CYCLE

Nitrogen is a key limiting factor for plant growth and development. N requirements differ greatly between plant species. Insufficient N uptake often leads to dwarf growth, ultimately resulting in plants with small cells and thickened cell walls, a condition known as sclerosis. In contrast, extreme supply of N, especially in the form of ammonia, results toxic to plants, causing reduced photosynthesis, displacement of cations and interference in hormone metabolism. In the long term, excessive N uptake by plants can favor sensitivity to abiotic and biotic stressors and delayed reproduction of the plant (Britto and Kronzucker 2002).

The main reserve of nitrogen in the biosphere is the molecular di-nitrogen (N_2) from the atmosphere, an inert gas with a triple bond that is energetically unfavorable to break. In this form, N cannot be directly utilized by plants, but needs to be first 'fixed' (reduced) to ammonia. A large number of species of free-living and symbiotic microbes, called diazotrophs, are able to carry out this process. When mediated by living organisms, the conversion of atmospheric dinitrogen to ammonia is also known as biological nitrogen fixation (BNF), and involves an enzyme called nitrogenase (nif) (Postgate 1998).

Although oxygen and respiration are generally crucial to maintain the energetic requirements associated with nitrogenase function, this enzyme is very susceptible to oxygen. This condition is also known as the "Oxygen Paradox". Several mechanisms have been developed by microorganisms to avoid the oxygen-mediated nitrogenase inactivation, which include: (i) ability to fix nitrogen anaerobically or in microaerobic conditions (Postgate 1998); (ii) increased levels of respiration to maintain the ATP production high for nitrogenase activity (Postgate 1998); and (iii) formation of specialized cellular structures (heterocyst) in which BNF takes place (Flores et al. 2015).

Among the members of the legumes (family *Leguminosae*), biological nitrogen fixation is particularly efficient, and the mechanism well characterized. The fixation process depends on a group of bacteria called rhizobia, mainly affiliated to α-proteobacteria (including the genera *Azorhizobium, Allorhizobium, Bradyrhizobium, Mesorhizobium, Rhizobium and Sinorhizobium*), and β-proteobacteria (Moulin et al. 2015). In this system, the nitrogen fixation occurs in the root- and stem nodules present on the leguminous plant. Nodules represent highly specialized structures in which plant-produced leghemoglobin binds and transfers oxygen to the rhizobia at concentrations harmless to their nitrogenase (Postgate 1998). The efficiency of the symbiotic relationship with nitrogen-fixing rhizobia provides legumes with an advantage over other plant species (Ferguson et al. 2010), making them particularly efficient for agricultural applications. Biological N transformation processes in soils are not only limited to fixation. Indeed, nitrification and denitrification are usually considered the dominant sources of nitric oxide (NO) production in soil, which in turn is the source of about 70% of the N2O emitted to the atmosphere (Conrad, 1996).

Nitrification is the biological oxidation of ammonia (NH_3) to nitrate (NO_3^-). This process is usually carried out by chemolithotrophic ammonia-oxidizers and nitrite-oxidizers. Members of β- and γ-Proteobacteria, such as *Nitrosomonas* spp., *Nitrosococcus* spp. and *Nitrosospira* spp. (Koops and Möller, 1992) are well characterized bacteria responsible for the oxidation of NH3, and their abundance is heavily influenced by soil pH and nitrogen addition (Levy-Booth et al. 2014).

Denitrification is a nitrate reduction process resulting in the production of molecular nitrogen (N_2) through a series of intermediate gaseous nitrogen oxide products, and is the primary pathway of N_2O emissions from soil (Colliver and Stephenson, 2000; Kowalchuk and Stephen, 2001; Shaw et al., 2006). Facultative anaerobic bacteria are mainly responsible for this process, which is carried out through the reduction of oxidized forms of nitrogen -mainly nitrate (NO_3−), nitrite (NO_2−), nitric oxide (NO), nitrous oxide (N_2O)- in response to the oxidation of an electron donor, such as organic matter, during microbial respiration (Medinets et al. 2015). Denitrification rates and end-products are usually affected by various edaphic variables, such as pH, O_2, organic C, temperature and moisture (Knowles, 1982; Tiedje, 1988; Brumme and Beese, 1992; Saad and Conrad, 1993;Thomsen et al., 1994; Bergaust et al., 2008, 2010).

Denitrification usually occurs under anoxic conditions, as when oxygen levels are depleted nitrate becomes the primary oxygen source for microbial respiration. Soil denitrifying microorganisms are abundant components of the bacterial community, accounting for 0.5-5% of the total population (Henry et al., 2006; Demanèche et al., 2009; Bru et al., 2011). Denitrification potential occurs in a wide range of both heterotrophic (e.g., *Pseudomonas stutzeri, Pseudomonas aeruginosa,* and *Paracoccus denitrificans*) (Carlson and Ingraham, 1983) and autotrophic (e.g., *Thiobacillus denitrificans*) bacteria (Baalsrud and Baalsrud, 1954). Often this trait is also coupled with N-fixation and ammonia-oxidation abilities in many strains (Levy-Booth et al. 2014). However, in the last three decades increasing attention has been warranted to aerobic denitrifiers, usually isolated from sludge, wastewater, and wastewater treatment systems, or in general in environments where fluctuating oxygen concentrations and reduced carbon are available (Ji et al. 2015). Aerobic denitrifiers mainly belong to α-, β- and γ-Proteobacteria.

MICROORGANISM AND THE PHOSPHOROUS CYCLE

Phosphorus (P) is an essential nutrient for both animals and plants. It is critical in cell development and has an important role for storing energy in the form of ATP (adenosine triphosphate), DNA and lipids (fats and oils). In the biosphere, phosphorus (P) is usually present in minimal concentration (0.09 wt%) and, unlike other critical ecosystem components such as nitrogen and carbon, this element has no stable atmospheric gas phase. Consequently, ecosystems usually depend on aqueous transfer of phosphorous (Plante 2007).

Generally, phosphorus is present at noticeable concentrations only in few minerals, with fluorapatite representing the main reservoir for this precious mineral. In soils, weathering is a crucial step for the release of P from minerals. Several biogeochemical processes are involved in the P release, including: (i) biochemical respiration, which releases CO_2 and increases acidity around degrading organic matter and root hairs, resulting in P-bearing minerals rapid dissolution (e.g. Schlesinger 1997); and (ii) P-containing mineral dissolution mediated by organic acid exudates from plant roots (Schlesinger 1997). After dissolution, P is usually incorporated into plant tissues, where it is converted into an organic form. Once taken up by plants, phosphate is actively processed and transformed into biomolecules essential for photosynthesis and metabolism. In order to increase the supply of P to roots, plants are able to produce and secrete enzymes, such as phosphatase, which aid in the release of bioavailable inorganic P from organic matter into soil pore spaces (Bucher 2007).

Both fungi and bacteria play critical roles in increasing the uptake of phosphorus by plants. Indeed, the beneficial effect on plant growth of the plant-microbe symbioses in the form of mycorrhizae (from the greek *myces* = fungus, *rhizo* = roots) and plant growth promoting rhizobacteria

(PGPR) has been widely recognized, and is mostly attributable to an increase in the uptake of nutrients, especially phosphorus (P) (Richardson, 2007). Mycorrhizal fungi and PGPR are able to improve soil phosphorus availability by solubilizing inorganic P or by the mineralization of organic phosphorus. The microbe-mediated intensification in the uptake of P occurs through various mechanisms, including: (i) exploration of larger soil volume, which decreases the area necessary for plant roots to reach the diffused P ions and simultaneously increases the surface area for absorption; (ii) faster movement of P into mycorrhizal hyphae, due to an increased affinity for P ions and, in parallel, a decrease of the minimum concentration required for P absorption; and (iii) solubilization of soil phosphorus, mediated by the release of organic acids, siderophores and phosphatase enzymes (Marschner, 2008).

Microorganisms able to solubilize inorganic phosphate are collectively termed "phosphate solubilizing microorganisms" (PSMs). Among the best *in vitro* characterized PSMs we include bacteria from the genera *Alcaligenes, Arthrobacter, Azotobacter, Bradyrhizobium, Bacillus, Burkholderia, Chromobacterium, Enterobacter, Erwinia, Escherichia, Flavobacterium, Micrococcus, Pantoea, Pseudomonas, Salmonella, Serratia, Streptomyces, and Thiobacillus* (Zhao and Lin 2001; Sindhu et al. 2009; Castagno et al. 2011; Azziz et al. 2012), as well as fungi belonging to the widespread genera *Aspergillus, Fusarium, Penicillium, Rhizopus,* and *Sclerotium* (Zhao and Lin 2001).

Together with solubilization, microbes also play important roles in P mineralization (Richardson et al. 2009). As mineralization of organic P often is dependent upon substrate solubilization and hydrolysis by phosphatase enzymes many PSMs also possess the ability to mineralize P. For example, common rhizosphere microorganisms such as *Bacillus* and *Pseudomonas* among the bacteria, and *Aspergillus* and *Penicillium* among fungi, have been reported to have organic P hydrolytic abilities both *in vitro* and when added to soil. (Marschner, 2008). Soil microbes involved in P cycling can produce a wide range of enzymes, including non-specific ones, such as acid and alkaline phosphates, as well as specific ones, such as phytases, which catalyze the release of orthophosphate from phytate and other inositol phosphates.

Given the importance of microorganisms for supplying P to plants, microbes represent an attractive alternative to increase the availability of P in soil, especially in agricultural systems, where increased P availability can result in both enhanced crop production and sustainable agricultural system (Sanchez et al. 1997; Deubel and Merbach 2005; Richardson and Simpson 2011). The possibility to improve availability and plant uptake of phosphate under nutrient imbalance conditions by inoculation with PSMs with P mineralisation capabilities has been demonstrated in numerous studies (Glick 1995; Igual et al. 2001; Wu et al. 2005). Therefore, this approach holds promise to offer a greater agronomic advantage compared to the expensive inorganic sources of P fertilizers. Being able to exploit phosphate-solubilizing microorganisms could represent an important alternative for improving P nutrition in many natural ecosystems (Rodriguez and Fraga 1999; Bagyaraj et al. 2000; Richardson et al. 2001).

THE INTERPLAY BETWEEN MICROORGANISMS AND SOIL TEXTURE

Soil texture and soil structure are both important features that contributes to soil fertility and the ability of microorganisms to contribute to nutrient cycling. Soil texture impacts the moisture retention, air and water movement within a soil, nutrient binding and influences the distribution of microorganisms within the soil (Buscot *et al.* 2005). Soil texture is largely governed by the size of the mineral components within the soil. Soil particles can range from large, > 2 mm, particles, down to minute silt and clay particles (less than 2 μm; Table 1). The mixture of soil particles with diameters less than 2 mm (i.e. sand, silt and clay particles) determine the texture of a soil (Buscot *et al.* 2005).

Table 1: Size ranges of particle types that make up soil

Particle	Size (mm)
Gravel	> 2
Coarse sand	0.2 – 2
Fine sand	0.02 - 0. 2

Silt	0.002 – 0.02
Clay	< 0.002

Soils dominated by large particles generally have large pore sizes and are free draining but relatively poor nutrient binding ability. In contrast soil made up of small particles have a high nutrient and water holding capacity, but are prone to compaction, which restricts the movement of plant roots, and water logging (Brewer 1988).

Soil structure refers to the arrangement of soil particles into larger soil aggregates. Soil aggregates can be defined as "naturally occurring cluster of group of soil particles in which the forces holding the particle together are greater that the forces between adjacent particles" (Martin *et al.* 1955). Soil aggregation improves soil structure by increasing the soil pore space, reducing soil density and promoting root penetration and the exchange of air and water with plant roots (Bronick and Lal 2005). Fungi and bacteria can contribute to stabilization of soil aggregates through deposition of extracellular polysaccharides which can bind to soil particles and act as a glue for aggregate formation. Fungi also contribute to aggregate stabilization through hyphal anchoring of particles (Singer 2006). Conversely, the degree of soil aggregation can influence microbiological activity and is influenced by a range of abiotic factors including: soil organic carbon (SOC), clay, carbonates, and ionic reactions (Chotte, 2005; Bronick and Lal 2005).

CONCLUSIONS

Soil organic matter comprises only a small fraction of the soil by volume and consists of both living and dead organic matter. The living microbial component of SOM makes up just a fraction of the total soil volume but the activity of the microorganisms (bacteria and fungi) in the soil is critical for the decomposition of organic residues, such as decaying plant and animal material, and the recycling of important soil nutrients. Improving soil nutrient status is a critical aspect of ecological restoration thus incorporating our understanding of how microorganisms govern nutrient cycling processes is integral for a holistic approach to ecological restoration and remediation projects.

SELECTED REFERENCES

1. Augspurger, C. K. (1984). "Seedling survival of tropical tree species: interactions of dispersal distance, light-gaps, and pathogens." Ecology **65**(6): 1705-1712.
2. Blume, H.-P. (2016). Scheffer/schachtschabel soil science, Heidelberg Springer.
3. Brady, N. C. (2010). Elements of the nature and properties of soils. Upper Saddle River, N.J., Upper Saddle River, N.J.: Pearson Prentice Hall.
4. Brewer, R. (1988). The science of ecology. Philadelphia, Philadelphia: Saunders College Pub.
5. Bronick, C. J. and Lal, R. (2005). "Soil structure and management: a review." Geoderma **124**(1–2): 3-22.
6. Buscot, F., Varma, A. and Ebsco (2005). Microorganisms in soils: roles in genesis and functions. Berlin; New York, Berlin; New York : Springer.
7. Chotte, J.-L. (2005). Importance of Microorganisms for Soil Aggregation. Microorganisms in Soils: Roles in Genesis and Functions. A. Varma and F. Buscot. Berlin, Heidelberg, Springer Berlin Heidelberg**:** 107-119.
8. Dodds, W. K., Gudder, D. A. and Mollenhauer, D. (1995). "The ecology of *Nostoc*." Journal of Phycology **31**(1): 2-18.
9. Fierer, N., Bradford, M. A. and Jackson, R. B. (2007). "Toward an ecological classification of soil bacteria." Ecology **88**(6): 1354-1364.
10. Floudas, D., Binder, M., Riley, R., Barry, K., Blanchette, R. A., Henrissat, B., Hibbett, D. S. (2012). "The paleozoic origin of enzymatic lignin decomposition reconstructed from 31 fungal genomes." Science **336**(6089): 1715-1719.
11. Lugtenberg, B. and Kamilova, F. (2009). Plant-growth-promoting rhizobacteria. **63:** 541-556.
12. Martin, J. P., Martin, W. P., Page, J. B., Raney, W. A. and de Ment, J. D. (1955). Soil Aggregation. Advances in Agronomy. **7:** 1-37.
13. Novaes, E., Kirst, M., Chiang, V., Winter-Sederoff, H. and Sederoff,

R. (2010). "Lignin and biomass: A negative correlation for wood formation and lignin content in trees." Plant Physiology **154**(2): 555-561.

14. Raaijmakers, J. M., Paulitz, T. C., Steinberg, C., Alabouvette, C. and Moënne-Loccoz, Y. (2009). "The rhizosphere: A playground and battlefield for soilborne pathogens and beneficial microorganisms." Plant and Soil **321**(1-2): 341-361.
15. Singer, M. J. (2006). Soils: an introduction. Upper Saddle River, N.J., Upper Saddle River, N.J.: Pearson Prentice Hall.
16. Singh, B. K., Bardgett, R. D., Smith, P. and Reay, D. S. (2010). "Microorganisms and climate change: Terrestrial feedbacks and mitigation options." Nature Reviews Microbiology **8**(11): 779-790.
17. Sonkoly, J., Valkó, O., Deák, B., Miglécz, T., Tóth, K., Radócz, S., .Török, P. (2017). "A new aspect of grassland vegetation dynamics: cyanobacterium colonies affect establishment success of plants." Journal of Vegetation Science **28**(3): 475-483.
18. Thauer, R. K., Kaster, A. K., Seedorf, H., Buckel, W. and Hedderich, R. (2008). "Methanogenic archaea: Ecologically relevant differences in energy conservation." Nature Reviews Microbiology **6**(8): 579-591.
19. Van Der Heijden, M. G. A., Bardgett, R. D. and Van Straalen, N. M. (2008). "The unseen majority: Soil microbes as drivers of plant diversity and productivity in terrestrial ecosystems." Ecology Letters **11**(3): 296-310.

3

CHAPTER

THE ROLE OF MICROORGANISMS IN PLANT COMMUNITY DEVELOPMENT, SUCCESSION AND PLANT INVASION MECHANISMS

Eleonora Egidi[1], Jennifer Wood[2], Prashant Pant[3]

[1] Dr. Eleonora Egidi is a mycologist interested in different aspects of fungal research, including fungal ecology, biology and phylogeny. Eleonora has an international background and has been involved in collaborative projects for the identification of new fungal barcoding regions, study of fungal communities from extreme environments, and use of fungi in phytoremediation. Her current scientific interests focus on exploring the biodiversity of Australian fungal flora and their role as drives of plant diversity in natural ecosystems, as well as understanding how these systems respond to environmental changes.

[1]Jennifer Wood is post-doctoral fellow in Dr. Ashley Franks' Applied and Environmental Microbiology Laboratory at La Trobe University, Australia. She has recently completed her Doctoral research in Microbial Ecology at the same university. Her broad research interest is in understanding the ecology of soil-microbial communities and their interactions with macro-communities. She is interested in developing trait-based approaches to interpret and predict how environmental and anthropogenic changes impact microbial community ecology and function.

[3] Dr. Prashant Pant, is an Assistant Professor of Botany in Dyal Singh College, University of Delhi, New Delhi, India. He has been teaching for more than 5 years now and his main areas have been Bioinformatics, Molecular Biology and Genomics at the college. His PhD was on understanding the Microbial Succession using a Metagenomic approach in a Barren mined-out site undergoing ecological restoration. For this, he deployed many molecular and computational Biology tool and techniques. He has several publications

(articles and book chapters) in International Journals and International Publications. Currently, he is also Chief Editor of Bioinformatics Review and a co-founder/Executive Chief Editor of a Science Journal in Hindi Language "Swadeshi Vigyan" for the native language speaking communities. Both the Journals are a service to the scientific community as they are fully open access.

INTRODUCTION

The occurrence of a variety of biotic and abiotic factors (site or locality factors) involved in determining the structure and community diversity of plants has been widely recognized in ecological theory. Although their relative importance in shaping plant diversity over time and space is still debated, parameters such as light, nutrient competition, environmental stresses, disturbances, as well as edaphic and climate variables are all considered important drivers of diversity and dynamics in many plant ecosystems (Van der Putten et al. 1993; De Deyn et al. 2003, Reynolds et al. 2003). An increasing body of evidences has contributed to also highlight the role of plant-microbe interactions in shaping plant dynamics at the community level, stressing the importance of including microbial traits when assessing ecosystem health and remediation strategies (Reynolds et al. 2003, Dickie et al. 2017, Menzel et al. 2017). Rhizospheric microorganisms are predicted to be able to influence the composition and productivity of natural plant communities both directly and indirectly (van der Heijden et al., 1998, 2006, 2008; Schnitzer et al., 2011). Indeed, the importance of soil microbial diversity in influencing the aboveground diversity and productivity has been demonstrated in many situations and ecosystems (De Deyn et al., 2004; Hooper et al., 2005; van der Heijden et al., 2008; Lau and Lennon, 2011; Wagg et al., 2011). In this scenario, belowground diversity is also considered an important predictor of the ability of the plant community to maintain plant productivity under different environmental stressing conditions (Wagg et al., 2011). In this chapter, we will review the biotic factors involved in shaping patterns of diversity and structure in natural vegetation communities, with particular emphasis on the role of below ground microbes in regulating plant diversity, as well as competition dynamics, invasion success and resilience to disturbance at the plant community level.

ROLE OF MICROORGANISMS IN SHAPING PLANT FITNESS

Soil contains a large diversity of microorganisms, with the rhizospheric compartment accommodating many of the soil biota directly involved in determining the fitness of individual plants and, consequently, their fate within the community (Reynolds et al. 2003; Fierer, 2017). Indeed, plants are colonized by a vast number of microorganisms, whose effect on seed germination, seedling vigor, plant growth and development, nutrition, diseases, and productivity has been largely demonstrated (see for example Mendes et al. 2013). As outlined in chapter 2, the plant-associated microbial communities, often referred to as plant microbiome, provide the plant with specific functions, services and traits. In return, plants feed the microbial community by releasing their photosynthetically fixed carbon into their direct surroundings (Nelson, 2004; Frey-Klett et al., 2007; Raaijmakers et al., 2009; Berendsen et al., 2012; Vorholt, 2012), therefore determining the rhizosphere-associated microbial community structure, composition and activities.

The rhizosphere, defined as the narrow area at the plant root-soil interface, is considered as one of the most complex ecosystems, being a hot spot for the occurrence of many microorganisms (Hinsinger and Marschner, 2006; Pierret et al., 2007; Hinsinger and Courchesne 2008; Hinsinger et al., 2009; Raaijmakers et al., 2009). Together with nematodes, protozoa, algae, viruses, and arthropods, microbes found in the rhizospheric soil include bacteria, fungi, oomycetes, and archaea (Mendes et al. 2013). Most of the microorganisms normally found in this plant compartment take also part in a complex food web, that is mainly driven by the plant rhizodeposits. Indeed, the decline in microbial diversity sequentially observed from bulk soil to rhizosphere, rhizoplane, and roots in some plant species, such as in the model species *Arabidopsis thaliana* (Reinhold-Hurek et al. 2015), suggests an increasingly stronger competition among microorganisms in this compartment, with some of the microorganisms being preferentially promoted or inhibited. This mechanism is mainly driven by root exudates, which release large quantity of nutrients in the rhizosphere immediate surroundings, with important repercussions on the diversity of microbes and their resulting activity on plant roots.

Among the best characterized rhizosphere microorganisms, nitrogen-fixing bacteria, mycorrhizal fungi, plant growth-promoting rhizobacteria (PGPR), biocontrol microorganisms, and mycoparasitic fungi are well known to exert a beneficial influence on plant growth and productivity. In particular, symbiotic associations of both legumes with nitrogen-fixing rhizobia and many mycorrhizal fungi provide plants with access to important limiting nutrients, such as nitrogen and phosphorous, respectively (reviewed in chapter 2). Other rhizosphere colonizers are known to be able to mobilize nutrients that are not readily available to plants, such as phosphorous, nitrogen or iron, via mechanisms such as solubilization, mineralization, or excretion of siderophores (Mendes et al. 2013). In addition, several rhizosphere bacteria produce and secrete an array of compounds that regulate plant growth, including auxins, cytokinins and gibberellins (Badri et al 2009, García de Salamone et al 2001, Rani et al 2011). Auxin (indole-3-acetic acid, IAA) in particular is the major plant hormone that controls apical dominance, phototropism, shoot elongation, as well as lateral and adventitious root growth (Elobeid et al 2012).

In addition to providing plants with essential nutrients and mediating the release of growth hormones, the ability of plant microbiota to increase plant tolerance to abiotic stresses, including fire, drought, flooding, osmotic stress, extreme temperatures, and heavy metal contamination has been widely reported in literature. For example, many plant-growth-promotion processes have been implicated in improving the rate of heavy metal hyper-accumulation either by increasing plant biomass or reducing plant stress. 1-aminocyclopropane-1-carboxylic acid (ACC) deaminase production is one of the main mechanisms through which rhizobacteria are able to alleviate plant stress. Indeed, ACC is a precursor to the plant stress signaling hormone, ethylene. It is produced by plants under stressful conditions, including heavy metal contamination (Glick 2003, Onofre-Lemus et al., 2009). Microbial enzymes that catabolize ACC to α-ketobutyrate and ammonia, result in the ACC removal from the rhizosphere and create a chemical gradient that perpetuates further ACC efflux, decreasing the concentration of ACC in plant tissues (Onofre-Lemus et al., 2009). The removal of ACC inhibits ethylene production and normal plant stress responses including conservation of root exudates and decreased root proliferation (Burd et al., 1998, Ma et al., 2009). Similarly, fungi are important players in alleviating the plant stress in hydrocarbons polluted soils, as was shown by inoculation of the endophytic fungus *Lewia* sp. in the rhizosphere of *Festuca arundinacea* (Cruz-Hernández et al., 2013).

Microorganisms can also negatively affect the plant fitness. For example, soil borne pathogens, which are usually found to grow and survive in the bulk soil, can establish a parasitic relationship with the host plant in the rhizosphere compartment. Most of soil borne pathogens belong to nematodes, oomycetes and fungi, while relatively few bacteria are pathogenic. Indeed, bacteria usually require a wound or natural opening to penetrate the plant host and cause infection. On the contrary, thanks to their ability to kill and penetrate host tissue with enzymes and toxins produced by the hyphae, Fungi and oomycetes are particularly efficient pathogenic agents for plants. Some pathogens, including rusts and powdery mildews, are found to occur on the above-ground compartments of the plants and require a living cell to obtain nutrients, and are called biotrophic pathogens. Several root pathogens, such as *Phytophthora* spp., are semi-biotrophic, while surprisingly few root pathogens, such as zoosporic fungi and Oomycetes, are found to be biotrophic (Raaijmakers et al. 2009).

The complex rhizosphere community interacts with the soil-borne pathogen, with the composition of the community being influenced by the pathogen infection. These interactions have important implications for the plant fitness, growth and survival. For example, a disease-suppressive sugar beet rhizosphere community was found to be strongly affected by the presence of the pathogenic fungus *Rhizoctonia solani*, resulting in an increase of stress-resistant bacteria in the plant-associated microbial community (Chapelle et al. 2016). Similar shifts in the rhizosphere community were reported following *Rhizoctonia solani* infection in the phyllosphere of maize (Manching et al. 2014) and in the lettuce rhizosphere and phyllosphere (Chowdhury et al. 2014), which caused severe symptoms and reduced growth in the affected plants. These results suggest that inter-microbial interactions can strongly influence the rhizosphere community composition and, thus, the host plant health (Chapelle et al. 2016).

PLANT-SOIL FEEDBACK

As described above, plant-associated microbial communities can both suppress plant health and fitness, and alleviate biotic and abiotic stresses on plants under various adverse conditions. The predominant role of rhizosphere biota in influencing plant growth can be extended also at the plant community level. Indeed, soil microbes can drive the structure and diversity of plant communities by either specifically reducing dominant plant species (Van der Putten et al. 1993; De Deyn et al. 2003) or by enhancing subordinate plant species (Van der Heijden et al. 1998). The impact of soil biota on plant survival and growth, and their role in maintaining plant diversity, is therefore predicted to depend on traits such as plant nutrient-acquisition strategy and microbial pathogenicity (Laliberte et al. 2015).

These interactions between plants and their biotic and abiotic soil environment are called plant–soil feedback (Bever et al. 1997; Ehrenfeld et al. 2005). The dynamics of the soil-plant synergism can be distinguished in positive or negative, depending on the outcome of the direct effects of the microbial community on the individual plant types (Klironomos 2002, Reynolds et al. 2003). Given the complexity and various strength degrees of these synergisms, the plant community succession will therefore depend mainly on the direction and magnitude of plant–soil feedbacks, which have been predicted to greatly vary along environmental gradients (Reynolds et al. 2003).

In the case of a positive feedback, the accumulation of microbes beneficial for a particular plant species reinforces the abundance of the associated host, resulting in the loss of other plant types from the local community, at least on a local scale (Bever et al. 1997, Bever 1999). For example, the occurrence of microbes such as arbuscular mycorrhizal fungi or PGPR can promote dominant plant species (Hartnett and Wilson 1999), while soil pathogens may progressively cause a decrease of the rare plants (Klironomos 2002), resulting in a decline in local plant species diversity via conspecific negative density dependence. This process might be particularly detrimental for non-mycorrhizal plant species, characterized by less well defended roots (Laliberté et al. 2015) in comparison to arbuscular and ectomycorrhizal plants, whose better protection against pathogens could prevent pathogen-mediated negative density dependence and even promote mono-dominance via positive plant-soil feedback (Newbery et al. 2000).

When a negative feedback occurs, a specific plant performs relatively worse within its community because of a change in the native microbial community. For example, the accumulation of soil borne pathogens or host-specific changes in the composition of the soil microorganisms can result in an improved growth of another plant species, consequently to an indirect facilitation mediated by the change in the microbial community (Reynolds et al. 2003). In boreal temperate forests, negative feedbacks have been found to drive the reciprocal replacement and coexistence of the main dominant species (Fox 1977, Woods 1979), while the high mortality of black cherry (*Prunus serotina*) seedlings was found to be associated with the accumulation of pathogenic *Pythium* spp. in high seedling density situations (Packer and Clay 2000). Similarly, in grassland communities, graminoids and small herbs have been found to undergo a negative feedback in response to their own soil biota, in contrast with tall herbs that showed a positive response, while legumes responded neutrally (Cortois et al. 2016). In the study by Cortois and colleagues (2016), all grassland plants benefitted from growing with other species' soil microorganisms, indicating an increased level of specialization in plant pathogens in comparison with growth-promoting soil biota. Moreover, species with lower degree of mycorrhizal colonization experienced negative soil feedback when grown in not-indigenous soil, suggesting that nutrient-acquisition strategy could be a useful trait to generalize and predict plant responses to soil biota in natural, mixed-species communities (Cortois et al. 2016, Teste et al. 2017).

PLANT-SOIL FEEDBACK AND PLANT INVASION MECHANISMS

Along the parasitic-mutualistic continuum, both pathogenic and mutualistic microbes are predicted to strongly influence many plant-microorganism symbiotic interactions. Both positive and negative dynamics are able to drive downstream effects, including trophic relationships, loss of biodiversity, modification of carbon and nutrient dynamics, and ecosystem resilience (Dickie et al., 2011; Corbin &

D'Antonio, 2004). These synergisms are especially relevant in defining the ecological and evolutionary impacts of invasive plant species (Mitchell et al., 2006; Pringle et al., 2009), as microbial symbionts introduced together with non-native plant species can critically influence the colonization patterns and invasion outcome in the new habitat.

Many hypotheses that have been proposed to frame the ecological outcomes linked to biological invasions, such as the Enemy Release Hypothesis (ERH), the Novel Weapons Theory, and the Biotic Resistance Hypothesis, can be successfully applied to the linked pathogen-plant invasion. For example, as recently summarized in an excellent review from Amsellem and colleagues (2017), the loss of microbial enemies often follows successful invasion, since many pathogenic microorganisms are not carried out in the newly colonized habitat by the introduced plants, resulting in increased plant fitness and promoted invasion (Mitchell and Power, 2003; Rúa et al., 2011; van Kleunen and Fischer, 2009). However, novel detrimental host–microorganism associations can be formed following invasion, as a consequence of pathogen communities recruitment by introduced hosts (e.g. Mei et al., 2014), or occurrence of new pathogenic microorganisms -both native and/or alien- resulting from the loss of microorganism enemies during invasion (Stricker et al., 2016).

Together with the investigation of biological outcomes associated with the enemy release, an increasing body of research is also focusing on unraveling the correlation between mycorrhizal symbiosis and plant invasiveness, drawing the attention of invasion ecologists on the role of mycorrhizal fungi in determining plant invasion success (Dickie et al. 2017). Indeed, increasing evidence suggests that maintenance or reestablishment of fungal symbioses commonly occur during plant invasions (see for example Menzel et al. 2017).

Recently, Dickie et al. (2017) have implemented a motif approach, adopted from network theory, to frame the complexity of invasive plant-mycorrhizal dynamics into ecological hypotheses able to predict the invasion dynamics. In their framework, the maintenance of vital mutualistic interactions may occur either through co-introduction and co-invasion, or when alien plant-associated symbionts are already present as natives in the invaded habitat. Co-introduction is usually the result of transport of infected plants or propagules, or a consequence of the transport of soil together with the invasive plants (Rudgers et al., 2005; Shipunov et al., 2008). Following co-introduction, the symbiotic partner not only can facilitate the invasion, but it can become itself invasive, as observed in the case of pines with their native fungal symbionts in the Southern Hemisphere (Dickie et al., 2010; Hayward et al., 2015). In alternative, the symbiont may be already present in the invasive range. This is the case, for example, of arbuscular mycorrhizal fungi, which show low endemism (Davison et al., 2015) and low specificity in their range of associations, therefore increasing the probability of alien plants to find a suitable mycorrhizal partner in the invaded habitat.

A third possibility is represented by the ability of the invasive plant to form novel symbiotic associations in the invaded range. A well-documented case is represented by the association of many Australian *Eucalyptus* spp. with the ectomycorrhizal fungus *Amanita muscaria* following the European colonization (Nunez and Dickie, 2014). Remarkably, when the novel association involves a native pathogen and an alien plant, the negative outcome of this interaction may be beneficial for the invaded community, resulting in biotic resistance, and thus prevented invasion (Levine et al., 2004). It is interesting to note that in ecological restoration and conservation, the co-introduction of native pathogens can be exploited as a strategy to control alien plant species invasion. An example of the successful application of pathogenic fungi as biocontrol agents is represented by the use of *Phragmidium violaceum* to control the spread of invasive blackberry in Australia (Morin and Evans, 2012), Chile (Oehrens 1977) and USA (Johnson and Mahaffee 2010).

At the ecosystem level, the linked plant–fungal invasions can result in dramatic changes in function and processes, which have been summarised by Dickie and colleagues (2017). For example, the invasion of non-ectomycorrhizal communities by ectomycorrhizal plants results in both the decrease of soil carbon and the co-release of nutrients, such as phosphorous and nitrogen, mediated by the fungal enzymatic activity (Dickie et al., 2014). In addition, the plant–fungus co-invasion may result in changes in total plant biomass, either increased by the mutualist fungus (Dickie et al., 2011) or decreased by the pathogenic invader (Mitchell, 2003; Cobb & Rizzo, 2016; Preston et al., 2016). Both the shifts in biomass and the plant compositional changes represent critical determinants of change in ecosystem processes (Ehrenfeld, 2010; Cobb & Rizzo, 2016; Preston et al., 2016).

CONCLUSIONS

In this chapter, we summarized the role of plant-associated microbial communities in determining plant community development, succession, and plant invasion mechanisms. Changes in soil microbial communities can positively or negatively influence plant health and fitness, determining the structure and dynamics of plant communities by either specifically suppressing dominant plant species or increasing subordinate plant species. Moreover, microbial symbionts (both pathogens and mutualists) introduced together with non-native plant species can dramatically impact the community dynamic in the invaded range, with important repercussions for many ecosystem processes. When a native plant associates with an alien pathogen, long-term compositional changes may occur, with length and effect being strongly dependent on the traits of the species that are lost and the traits of the species that are gained. Therefore, we predict that understanding the dependence of plants on their associated microbes may be a key approach to design efficient restoration strategies, either focused to promote native plants biodiversity or managing invasive plants.

SELECTED REFERENCES

1. Amsellem, L., Brouat, C., Duron, O., Porter, S. S., Vilcinskas, A., & Facon, B. Importance of microorganisms to macroorganisms invasions— Is the Essential Invisible to the Eye? (The Little Prince, A. de Saint-Exupéry, 1943)
2. Badri, D. V., & Vivanco, J. M. (2009). Regulation and function of root exudates. *Plant, Cell & Environment*, *32*(6), 666-681.
3. Berendsen, R. L., Pieterse, C. M., & Bakker, P. A. (2012). The rhizosphere microbiome and plant health. *Trends in plant science*, *17*(8), 478-486.
4. Bever, J. D., Westover, K. M., & Antonovics, J. (1997). Incorporating the soil community into plant population dynamics: the utility of the

feedback approach. *Journal of Ecology*, 561-573.

5. Bever, J. D. (1999). Dynamics within mutualism and the maintenance of diversity: inference from a model of interguild frequency dependence. *Ecology Letters*, *2*(1), 52-61.
6. Burd, G. I., Dixon, D. G., & Glick, B. R. (1998). A plant growth-promoting bacterium that decreases nickel toxicity in seedlings. *Applied and Environmental Microbiology*, *64*(10), 3663-3668.
7. Chapelle, E., Mendes, R., Bakker, P. A. H., & Raaijmakers, J. M. (2016). Fungal invasion of the rhizosphere microbiome. *The ISME journal*, *10*(1), 265.
8. Chowdhury, S. P., Uhl, J., Grosch, R., Alquéres, S., Pittroff, S., Dietel, K., ... & Hartmann, A. (2015). Cyclic lipopeptides of Bacillus amyloliquefaciens subsp. plantarum colonizing the lettuce rhizosphere enhance plant defense responses toward the bottom rot pathogen Rhizoctonia solani. *Molecular Plant-Microbe Interactions*, *28*(9), 984-995.
9. Corbin, J. D., & D'Antonio, C. M. (2004). Competition between native perennial and exotic annual grasses: implications for an historical invasion. *Ecology*, *85*(5), 1273-1283.
10. Cortois, R., Veen, G. F., Duyts, H., Abbas, M., Strecker, T., Kostenko, O., ... & Putten, W. H. (2017). Possible mechanisms underlying abundance and diversity responses of nematode communities to plant diversity. *Ecosphere*, *8*(5).
11. Cruz-Hernández, A., Tomasini-Campocosio, A., Pérez-Flores, L. J., Fernández-Perrino, F. J., & Gutiérrez-Rojas, M. (2013). Inoculation of seed-borne fungus in the rhizosphere of Festuca arundinacea promotes hydrocarbon removal and pyrene accumulation in roots. *Plant and soil*, *362*(1-2), 261-270.
12. Dickie, I. A., & Holdaway, R. J. (2011). Podocarp roots, mycorrhizas, and nodules. *Smithsonian Contributions to Botany*, (95).
13. Dickie, I. A., Bufford, J. L., Cobb, R. C., Desprez-Loustau, M. L., Grelet, G., Hulme, P. E., ... & Thrall, P. H. (2017). The emerging science of linked plant–fungal invasions. *New Phytologist*.
14. De Deyn, G. B., Raaijmakers, C. E., & Van der Putten, W. H. (2004). Plant community development is affected by nutrients and soil biota. *Journal of Ecology*, *92*(5), 824-834.

15. Elobeid, M., & Polle, A. (2012). Interference of heavy metal toxicity with auxin physiology. In *Metal Toxicity in Plants: Perception, Signaling and Remediation* (pp. 249-259). Springer Berlin Heidelberg.

16. Ehrenfeld, J. G., Ravit, B., & Elgersma, K. (2005). Feedback in the plant-soil system. *Annu. Rev. Environ. Resour.*, *30*, 75-115.

17. Fox, J. F. (1977). Alternation and coexistence of tree species. *The American Naturalist*, *111*(977), 69-89.

18. Frey-Klett, P., Garbaye, J. A., & Tarkka, M. (2007). The mycorrhiza helper bacteria revisited. *New phytologist*, *176*(1), 22-36.

19. García de Salamone, I. E., Hynes, R. K., & Nelson, L. M. (2001). Cytokinin production by plant growth promoting rhizobacteria and selected mutants. *Canadian Journal of microbiology*, *47*(5), 404-411.

20. Glick, B. R. (2003). Phytoremediation: synergistic use of plants and bacteria to clean up the environment. *Biotechnology advances*, *21*(5), 383-393.

21. Kardol, P., Martijn Bezemer, T., & Van Der Putten, W. H. (2006). Temporal variation in plant–soil feedback controls succession. *Ecology Letters*, *9*(9), 1080-1088.

22. Klironomos, J. N. (2002). Feedback with soil biota contributes to plant rarity and invasiveness in communities. *Nature*, *417*(6884), 67-70.

23. Hartnett, D. C., & Wilson, G. W. (1999). Mycorrhizae influence plant community structure and diversity in tallgrass prairie. *Ecology*, *80*(4), 1187-1195.

24. Hinsinger, P., & Marschner, P. (2006). Rhizosphere—perspectives and challenges—a tribute to Lorenz Hiltner 12–17 September 2004—Munich, Germany. *Plant and Soil*, *283*(1), vii-viii.

25. Hinsinger, P., & Courchesne, F. (2008). *Biogeochemistry of metals and metalloids at the soil-root interface*. John Wiley & Sons: Hoboken, NJ.

26. Hinsinger, P., Bengough, A. G., Vetterlein, D., & Young, I. M. (2009). Rhizosphere: biophysics, biogeochemistry and ecological relevance. *Plant and soil*, *321*(1-2), 117-152.

27. Hooper, D. U., Chapin, F. S., Ewel, J. J., Hector, A., Inchausti, P., Lavorel, S., ... & Schmid, B. (2005). Effects of biodiversity on ecosystem functioning: a consensus of current knowledge. *Ecological monographs*, *75*(1), 3-35.
28. Laliberté, E., Lambers, H., Burgess, T. I., & Wright, S. J. (2015). Phosphorus limitation, soil-borne pathogens and the coexistence of plant species in hyperdiverse forests and shrublands. *New Phytologist*, *206*(2), 507-521.
29. Lau, J. A., & Lennon, J. T. (2011). Evolutionary ecology of plant–microbe interactions: soil microbial structure alters selection on plant traits. *New Phytologist*, *192*(1), 215-224.
30. Ma, Y., Rajkumar, M., & Freitas, H. (2009). Improvement of plant growth and nickel uptake by nickel resistant-plant-growth promoting bacteria. *Journal of Hazardous Materials*, *166*(2), 1154-1161.
31. Manching, H. C., Balint-Kurti, P. J., & Stapleton, A. E. (2014). Southern leaf blight disease severity is correlated with decreased maize leaf epiphytic bacterial species richness and the phyllosphere bacterial diversity decline is enhanced by nitrogen fertilization. *Frontiers in plant science*, *5*.
32. Mei, L., Zhu, M., Zhang, D. Z., Wang, Y. Z., Guo, J., & Zhang, H. B. (2014). Geographical and temporal changes of foliar fungal endophytes associated with the invasive plant Ageratina adenophora. *Microbial ecology*, *67*(2), 402-409.
33. Mendes, R., Garbeva, P., & Raaijmakers, J. M. (2013). The rhizosphere microbiome: significance of plant beneficial, plant pathogenic, and human pathogenic microorganisms. *FEMS microbiology reviews*, *37*(5), 634-663.
34. Menzel, A., Hempel, S., Klotz, S., Moora, M., Pyšek, P., Rillig, M. C., ... & Kühn, I. (2017). Mycorrhizal status helps explain invasion success of alien plant species. *Ecology*, *98*(1), 92-102.
35. Mitchell, C. E., Agrawal, A. A., Bever, J. D., Gilbert, G. S., Hufbauer, R. A., Klironomos, J. N., ... & Seabloom, E. W. (2006). Biotic interactions and plant invasions. *Ecology letters*, *9*(6), 726-740.
36. Mitchell, C. E., & Power, A. G. (2003). Release of invasive plants from fungal and viral pathogens. *Nature*, *421*(6923), 625.

37. Nelson, L. M. (2004). Plant growth promoting rhizobacteria (PGPR): Prospects for new inoculants. *Crop Management*, *3*(1), 0-0.

38. Newbery, D. M., Alexander, I. J., & Rother, J. A. (2000). Does proximity to conspecific adults influence the establishment of ectomycorrhizal trees in rain forest?. *The New Phytologist*, *147*(2), 401-409.

39. Onofre-Lemus, J., Hernández-Lucas, I., Girard, L., & Caballero-Mellado, J. (2009). ACC (1-aminocyclopropane-1-carboxylate) deaminase activity, a widespread trait in Burkholderia species, and its growth-promoting effect on tomato plants. *Applied and Environmental Microbiology*, *75*(20), 6581-6590.

40. Pierret, A., Doussan, C., Capowiez, Y., & Bastardie, F. (2007). Root functional architecture: a framework for modeling the interplay between roots and soil. *Vadose Zone Journal*, *6*(2), 269-281.

41. Pringle, A., Bever, J. D., Gardes, M., Parrent, J. L., Rillig, M. C., & Klironomos, J. N. (2009). Mycorrhizal symbioses and plant invasions. *Annual Review of Ecology, Evolution, and Systematics*, *40*, 699-715.

42. Raaijmakers, J. M., Paulitz, T. C., Steinberg, C., Alabouvette, C., & Moënne-Loccoz, Y. (2009). The rhizosphere: a playground and battlefield for soilborne pathogens and beneficial microorganisms. *Plant and soil*, *321*(1-2), 341-361.

43. Rani, M. U., & Reddy, G. (2011). Bacillus cereus and Enterobacter cancerogenus screened for their efficient plant growth promoting traits rhizobacteria (PGPR) and antagonistic traits among sixteen bacterial isolates from rhizospheric soils of pigeon pea. *African Journal of Microbiology Research*, *5*(15), 2090-2094.

44. Reynolds, H. L., Packer, A., Bever, J. D., & Clay, K. (2003). Grassroots ecology: plant–microbe–soil interactions as drivers of plant community structure and dynamics. *Ecology*, *84*(9), 2281-2291.

45. Reinhold-Hurek, B., Bünger, W., Burbano, C. S., Sabale, M., & Hurek, T. (2015). Roots shaping their microbiome: global hotspots for microbial activity. *Annual review of phytopathology*, *53*, 403-424.

46. Rúa, M. A., Pollina, E. C., Power, A. G., & Mitchell, C. E. (2011).

The role of viruses in biological invasions: friend or foe?. *Current opinion in virology*, *1*(1), 68-72.

47. Rudgers, J. A., Mattingly, W. B., & Koslow, J. M. (2005). Mutualistic fungus promotes plant invasion into diverse communities. *Oecologia*, *144*(3), 463-471.
48. Schnitzer, S. A., Klironomos, J. N., HilleRisLambers, J., Kinkel, L. L., Reich, P. B., Xiao, K., ... & Van Nes, E. H. (2011). Soil microbes drive the classic plant diversity–productivity pattern. *Ecology*, *92*(2), 296-303.
49. Stricker, K. B., Harmon, P. F., Goss, E. M., Clay, K., & Luke Flory, S. (2016). Emergence and accumulation of novel pathogens suppress an invasive species. *Ecology letters*, *19*(4), 469-477.
50. Teste, F. P., Kardol, P., Turner, B. L., Wardle, D. A., Zemunik, G., Renton, M., & Laliberté, E. (2017). Plant-soil feedback and the maintenance of diversity in Mediterranean-climate shrublands. *Science*, *355*(6321), 173-176.
51. Van der Heijden, M. G., Klironomos, J. N., Ursic, M., & Moutoglis, P. (1998). Mycorrhizal fungal diversity determines plant biodiversity, ecosystem variability and productivity. *Nature*, *396*(6706), 69.
52. Van Der Heijden, M. G., Streitwolf-Engel, R., Riedl, R., Siegrist, S., Neudecker, A., Ineichen, K., ... & Sanders, I. R. (2006). The mycorrhizal contribution to plant productivity, plant nutrition and soil structure in experimental grassland. *New Phytologist*, *172*(4), 739-752.
53. Van Der Heijden, M. G., Bardgett, R. D., & Van Straalen, N. M. (2008). The unseen majority: soil microbes as drivers of plant diversity and productivity in terrestrial ecosystems. *Ecology letters*, *11*(3), 296-310.
54. Van der Putten, W. H., Van Dijk, C., & Peters, B. A. M. (1993). Plant-specific soil-borne diseases contribute to succession in foredune vegetation. *Nature*, *362*(6415), 53-56.
55. Van Kleunen, M., & Fischer, M. (2009). Release from foliar and floral fungal pathogen species does not explain the geographic spread of naturalized North American plants in Europe. *Journal of Ecology*, *97*(3), 385-392.
56. Vorholt, J. A. (2012). Microbial life in the phyllosphere. *Nature*

reviews. Microbiology, *10*(12), 828.

57. Wagg, C., Jansa, J., Schmid, B., & van der Heijden, M. G. (2011). Belowground biodiversity effects of plant symbionts support aboveground productivity. *Ecology Letters*, *14*(10), 1001-1009.
58. Woods, K. D. (1979). Reciprocal replacement and the maintenance of codominance in a beech-maple forest. *Oikos*, 31-39.
59. Fierer, N., (2017). Embracing the unknown: disentangling the complexities of the soil microbiome. *Nature Reviews Microbiology*, doi:10.1038/nrmicro.2017.87 Published online 21 August 2017

4

CHAPTER

MICROBIAL AMELIORATION AND REMEDIATION OF ENVIRONMENTAL CONTAMINANTS

Jennifer Wood[1], Eleonora Egidi[2], Prashant Pant[3]

[1]Jennifer Wood is post-doctoral fellow in Dr. Ashley Franks' Applied and Environmental Microbiology Laboratory at La Trobe University, Australia. She has recently completed her Doctoral research in Microbial Ecology at the same university. Her broad research interest is in understanding the ecology of soil-microbial communities and their interactions with macro-communities. She is interested in developing trait-based approaches to interpret and predict how environmental and anthropogenic changes impact microbial community ecology and function.

[2] Dr. Eleonora Egidi is a mycologist interested in different aspects of fungal research, including fungal ecology, biology and phylogeny. Eleonora has an international background and has been involved in collaborative projects for the identification of new fungal barcoding regions, study of fungal communities from extreme environments, and use of fungi in phytoremediation. Her current scientific interests focus on exploring the biodiversity of Australian fungal flora and their role as drives of plant diversity in natural ecosystems, as well as understanding how these systems respond to environmental changes.

[3] Dr. Prashant Pant, is an Assistant Professor of Botany in Dyal Singh College, University of Delhi, New Delhi, India. He has been teaching for more than 5 years now and his main areas have been Bioinformatics, Molecular Biology and Genomics at the college. His PhD was on understanding the Microbial Succession using a Metagenomic approach in a Barren mined-out site undergoing ecological restoration. For this, he deployed many molecular and computational Biology tool and techniques. He has several publications

(articles and book chapters) in International Journals and International Publications. Currently, he is also Chief Editor of Bioinformatics Review and a co-founder/Executive Chief Editor of a Science Journal in Hindi Language "Swadeshi Vigyan" for the native language speaking communities. Both the Journals are a service to the scientific community as they are fully open access.

INTRODUCTION

Soils are chemically and biologically complex systems that can absorb and retain a wide variety organic and inorganic contaminant (Table 1). Organic contaminants can include: polyaromatic hydrocarbons (PAHs), petroleum hydrocarbons (PHC), explosives residues such as trinitrotoluene (TNT), halogenated hydrocarbons, industrial solvents and pesticide residues such as atrazine (Glick 2003). Inorganic contaminants include heavy metals such as cadmium (Cd), lead (Pb) and copper (Cu); radioactive metals such as uranium (U), biologically toxic non-metals such as arsenic (As) and selenium (Se) and other inorganic compounds such as phosphate and nitrate (Wood *et al.* 2016). Soil contaminants can be introduced into the environment via an array of anthropological activities, prominent amongst which are mining activities (mine tailings, the transport of ores, smelting etc.), agricultural activities (the application of fertilizers, pesticides or herbicides which also contain heavy metal impurities), and industrial activities (Liu *et al.* 2005, Wasi *et al.* 2013).

Bioremediation refers to the use of biological agents, usually bacteria or plants, to remediate contaminated soils and water. Bioremediation can either be achieved via the bio-degradation, bio-transformation or bio-sequestration of environmental contaminants, with different types of contaminants lending themselves to different bioremediation strategies (Salt *et al.* 1998, Lovley *et al.* 2004, Haritash and Kaushik 2009). For example, phytoextraction, which utilizes plants to move heavy metals from soils into the aerial tissues of plants where they are sequestered, is the predominant bioremediation strategy investigated for the remediation of heavy metal contaminated soils, because heavy metals cannot be degraded (Singh *et al.* 2003). In contrast, biodegradation is the preferred strategy to remediate hydrocarbon contaminated soils as the microbial component can be directly stimulated to metabolise and degrade hydrocarbons, or other organic pollutants (Liu *et al.* 2010).

Table 1: List of the top 10 most hazardous substances that pose the greatest threat to human health according to The United States Agency for Toxic Substances and Disease Registry (ATSDR) as of 2015. Where multiple species of a toxic substances are listed * indicate the more toxic form

ATSDR rank	Substance name	Contaminant class	Chemical structure/ ionic species
1	Arsenic	Heavy metal(liod)	As^{5+} (arsenate)*, As^{3+} (arsenite)
2	Lead	Heavy metal	Pb^{2+}
3	Mercury	Heavy metal(liod)	Hg^0, Hg^{2+}, $H_3C\text{-}Hg^+X^-$ (methylmercury)*
4	Vinyl chloride	Organic chlorine compound	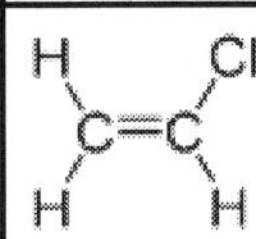
5	Polychlorinated biphenyls (general, 201)	Organic chlorine compound	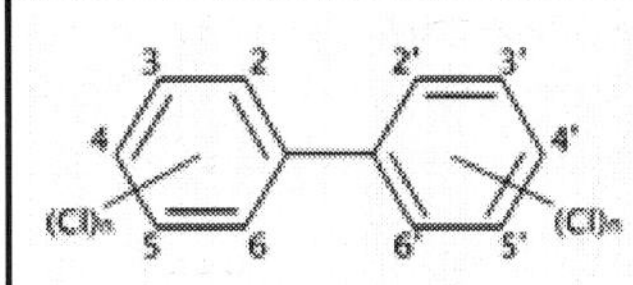
6	Benzene	Organic compound	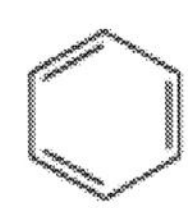
7	Cadmium	Heavy metal	Cd^{2+}
8	Benzo(a)pyrene	Polycyclic aromatic hydrocarbon	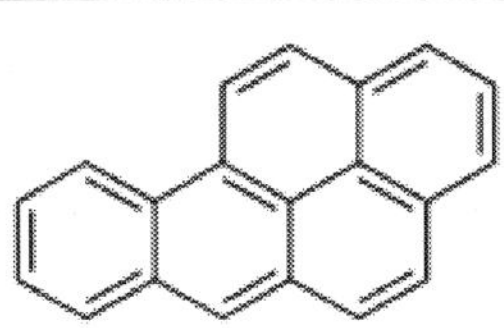

9	Polycyclic aromatic hydrocarbons (general, >100)	Polycyclic aromatic hydrocarbon	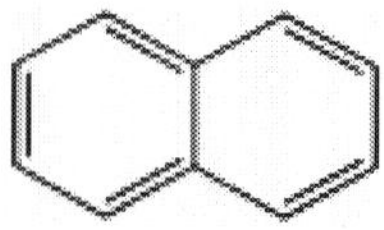 Naphthalene
10	Benzo(b) fluoranthene	Polycyclic aromatic hydrocarbon	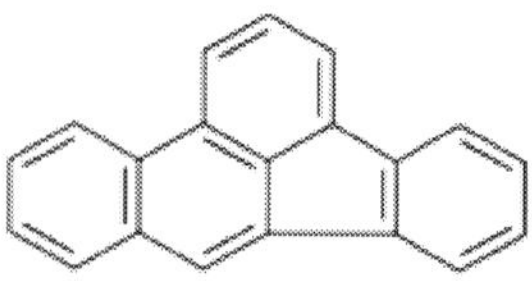

Interest in bioremediation strategies is intensifying as the rates at which contaminants are deposited into the environment, due to the increasing growth of industrial activities worldwide. An increasing body of research is now focusing on investigating the interplay between plants and soil microorganisms to improve contaminants remediation strategies. Indeed, soil microorganisms can play variety of roles in many bioremediation processes. Microorganisms can (i) act directly on the contaminants, degrading and detoxifying them as part of microbial metabolic processes (Carmona *et al.* 2009); (ii) transform contaminants, leading to their immobilization and detoxification or to their mobilization into plant tissues (Wood *et al.* 2016); (iii) indirectly improve bioremediation processes through positive feedback loops with plants involved in the bioremediation process (Wood *et al.* 2016).

In this chapter, we will review the role of microbial communities in many remediation processes, including phytoremediation and bioremediation. A better understanding of the interactions between plants and microorganisms is a fundamental task to disentangle the relative contributions, and synergistic effects, contributing towards contaminants remediation, opening for the possibility to develop more efficient remediation strategies.

MICROBIAL DEGRADATION OF SOIL CONTAMINANTS: HYDROCARBON CONTAMINANTS AND BIODEGRADATION

A priority amongst organic soil contaminants are polyaromatic hydrocarbons (PAHs). PAHs encompass a group of more than 100 chemicals all of which consist of a structure based on two or more fused benzene rings (Johnsen *et al.* 2005). Many PAH compounds, such as the pesticide naphthalene, have toxic, mutagenic and carcinogenetic properties (Haritash and Kaushik 2009). In addition to PAHs as a group, two specific PAHs, Benzo(a)pyrene and Benzo(b)fluoranthene are listed amongst The United States Agency for Toxic Substances and Disease Registry (ATSDR) top ten hazardous substances that pose the greatest threat to human health (Table 1; (Agency for Toxic Substances and Disease Registry 2015). PAHs are introduced into the environment due to the incomplete combustion of organic materials. These processes can occur naturally (e.g. wild fires, oil seeps, volcanic eruptions) or via anthropogenic activities (the burning of fossil fuels, coal, gas and wood; (Johnsen *et al.* 2005).

Microorganisms are metabolically diverse, capable of degrading a wide variety of substrates, and this metabolic diversity can be harnessed to remediate many organic contaminants, in a process known as bio-degradation. Many microorganisms have been isolated that have the potential for PAH bio-degradation. Amongst bacteria, the *Sphingomonas*, *Burkholderia*, *Pseudomonas, Rhodococcus* and *Mycobacterium* have been repeatedly shown to harbour PAH degrading strains (Kazunga and Aitken 2000) (Walter *et al.* 1991, Kästner *et al.* 1994, Schneider *et al.* 1996, Johnsen *et al.* 2005, Chebbi *et al.* 2017, Guo *et al.* 2017, Kong *et al.* 2017). Members of the genus *Sphingomonas*, in particular are known to degrade a wide range of toxic aromatic compounds, including pentchlorophenols, dioxins and PAHs, and represent a high proportion of the PAH-degrading isolates (Ederer *et al.* 1997, Halden *et al.* 1999, Johnsen *et al.* 2005, Aislabie and Deslippe 2013). *Sphingomonas* spp. are present in a wide variety of environments including soils and marine sediments. Many have been isolated as plant endophytes and found to have plant-growth-promoting attributes, suggesting that this group of bacteria are highly adapted to life in close association with plants (Chen *et al.* 2014, Khan *et al.* 2014, Sukweenadhi *et al.* 2015, Pan *et al.* 2016).

The ability of plant-associated bacteria to degrade complex aromatic contaminants is no coincidence. Root-associated (rhizosphere) microorganisms are adapted to detect and metabolize the diverse array of compounds that are secreted in plant-root exudates, many of which share structural similarities with manmade contaminants (Parales and

Harwood 2002). Plant derived aromatic compounds include: aromatic amino acids such as benzoate (Campbell *et al.* 1997, Carmona *et al.* 2009); terpenoid compounds such as limonene, pinene and geraniol (Lin *et al.* 2007, Owen *et al.* 2007); and other metabolites thought to be allelopathic, such as caffeine (Morita *et al.* 2011, Sugiyama *et al.* 2016). Man-made environmental contaminants with similar structural properties to aromatic root-exudes include polychlorinated biphenyls, dioxins and nitroaromatics (Carmona *et al.* 2009). These compounds have been shown to be elicit chemotactic effects and be degraded by microorganisms (Parales and Harwood 2002, Carmona *et al.* 2009). Thus, the metabolic diversity of rhizosphere-associated bacteria can be exploited to remediate contaminated soils.

Conversely, the rhizosphere can also be used to enrich for contaminant degrading microorganisms. Because the rhizosphere constitutes a stable, nutrient-rich environment it also fosters competitive interactions between microorganisms, as they compete for space and nutrients. In the presence of a soil contaminant, such as PAHs, contaminant-degrading abilities will confer a microorganism with a selective advantage. For instance, in the presence of hydrocarbons microorganisms capable of their degradation are selectively advantaged in soil (Hamamura *et al.* 2006). Interestingly, comparisons of microorganisms encoding hydrocarbon metabolising genes (*ndoB*, *alkB* and *xylE*) have demonstrated that, in the presence of contamination, the enrichment of hydrocarbon-degrading ecotypes is greater in the rhizosphere relative to bulk soils (Siciliano *et al.* 2001), indicating that selectively advantaged microorganisms in the soil are likely to be further enriched in the rhizosphere.

In addition to plant-bacteria interaction, plant-fungi interactions can also be harnessed for biodegradation. In particular the activity of lignoytic fungi (Haritash and Kaushik 2009). Lignin is a main structural polymer of plants, a highly complex molecule made up of cross-linked phenolic polymers that can be degraded by very few organisms (Parales and Harwood 2002, Novaes *et al.* 2010, Floudas *et al.* 2012). The only organisms capable of substantial lignin decay are the white rot fungi form the class Agricomycetes which use extracellular enzymes to degrade lignin including lignin peroxidases, Mn-dependent peroxidases and laccase (Johnsen *et al.* 2005, Floudas *et al.* 2012). Because the structure of lignin is irregular, the substrate specificity of lignin-degrading must necessarily be low in order to account for structural variations. This property imbues lignin degrading fungi with the ability to degrade a wide variety compounds with aromatic structures similar to that of lignin, including PAHs (Haritash and Kaushik 2009). Mixing fungal hyphae of the white-rot fungi *Pleurotus ostreatus* into PAH contaminated soil has been shown to improve the rate of 4- and 5-ring PAH degradation, under laboratory conditions (Eggen 1999). Similarly, the white rot fungus *Bjerkandera* sp. could degrade the highly mutagenic PAH, benzo[a] pyrene resulting in rapid and almost complete elimination of its high mutagenic potential (Kotterman *et al.* 1998).

A number of challenges still hamper the efficiency of bio-degradation processes. The chemistry of the contaminant can significantly alter its bioavailability for degradation. For example, being PAHs highly hydrophobic, they tend to interact with and bind to organic matter in the soil matrix. Indeed, the higher the molecular weight of the PAH, the more strongly it will sorbs to soils and sediments, becoming unavailable to microorganisms for bio-degradation (Mrozik *et al.* 2003). Additionally, although heterotrophic bacteria capable of degrading organic contaminants are ubiquitous in the soil, the toxicity of the soil contaminant itself can suppress microbial bio-degrading activity, resulting in the necessity of further stimulation for bio-degradation to proceed. For example, heavy petroleum contamination has been shown to suppress the growth of heterotrophic and PAH-degrading organisms in the soil (Liu *et al.* 2010). By simulating the native soil community with a 5% manure amendment, Liu and colleagues reported that the abundance of petroleum and PAH-degrading microorganisms increased 1 and 2-fold, respectively. The stimulation of the microbial community reduced total petroleum hydrocarbon of soil content by 58.2% in the treated soils compared with only 15.6% in unstimulated soils (Liu *et al.* 2010).

MICROBIAL TRANSFORMATION OF SOIL CONTAMINANTS: HEAVY METALS AND PHYTOEXTRACTION

Some soil contaminants cannot be degraded. In particular the accumulation of heavy metals in soil is of major concern as they are both toxic and persist within the environment, necessitating physical extraction or transformation into less toxic forms (Lum et al. 2014). Heavy metals can be defined as elements with a density greater than 5 g cm-3, but frequently the term is extended to refer more generally to elements associated with environmental contamination that threaten plant and animal health, resulting in the inclusion of some metalloids such as arsenic (As) and Selenium (Se) (Nagajyoti et al. 2010). The United States Agency for Toxic Substances and Disease Registry (ATSDR) has listed As, lead (Pb), mercury (Hg) and cadmium (Cd) amongst the top ten hazardous substances that pose the greatest threat to human health (Table 1). In humans, heavy metal poisoning can cause irreparable Although elevated levels of heavy metals in soils can occur naturally due the weathering of metalliferous rocks, a wide variety of anthropogenic activities can contribute to environmental heavy metal contamination, including: burning of fossil fuels, mining activities, municipal wastes, fertilizers and pesticides and sewage sludge amendments used in agriculture, and the use of pigments or batteries (Rajkumar et al. 2009, Nagajyoti et al. 2010). Heavy metals persisting in the soil can be absorbed by plant tissues, enter the biosphere and accumulate up the trophic levels of the food web (McLaughlin et al. 1999).

Several bioremediation strategies have emerged to remediate heavy metal contaminated soils. Phytoextraction is a low-cost, environmentally friendly remediation technique that utilizes plants to extract heavy metals from contaminated soil (Vamerali *et al.* 2010). In contrast to traditional heavy metal remediation strategies, which involve the physical removal of contaminated soil, chemical washing and reburial, it is estimated that phytoextraction could reduce operational costs as much as 30 fold and reduce environmental harm (Gerhardt *et al.* 2009). The interplay between plants and soil microorganisms is a highly studied research area aimed at optimizing this remediation strategy. In particular soil microorganisms can be used to manipulate the bioavailability of target metals, which is a key limiting factor for phytoextraction as it is for bio-degradation (Glick 2010).

Metal Reduction & Oxidation

Many soil microbes can influence the mobility and toxicity of heavy metals through oxidation or reduction (Gadd 2004). In particular dissimilatory metal-reducing bacteria, such as *Geobacter* species (Deltaproteobacteria), *Shewenella* species (Gammaproteobactreia) and *Deferribacter* (Deferribacteres), are well known for their ability to reduce a range of metals using organic compounds as electron donors (Gadd, 2004, Lovley *et al.* 2004). Depending on the oxidation state and the heavy metal in question, microbial redox reactions can increase or decrease heavy metal mobility (Bolan *et al.* 2013). For example, the reduction of U(VI) to U(IV), Tc(VII) to Tc(IV) or Tc(V), Co(III) to Co(II) and Cr(VI) to Cr(III) generate less toxic, insoluble ions (Kashefi and Lovley 2000). The metal-reducing capabilities of microbes like *Geobacter sulfurreducens* are being utilized in the remediation of uranium-contaminated groundwater through reduction-immobilization processes (Anderson *et al.* 2003, Zhuang *et al.* 2012). Of particular interest is the reported ability of *Geobacter* species to reduce and mobilize immobile As(V) and Cd to mobile As(III) (Muehe *et al.* 2013, Ohtsuka *et al.* 2013). For these metals, which are dangerous environmental contaminants, microbial reduction is a potential strategy to increasing the bioavailability and potentially phytoextraction rate of selected heavy metals (Stolz and Oremland 1999).

Organic Acid Production

In the plant rhizosphere, organic acids can increase heavy metal solubility via exchange reactions with heavy metals immobilized in complexes, or by dissolving metal-oxides that have formed on surfaces within the soil matrix. Some organic acids can also chelate heavy metals, forming soluble metal-organic complexes, and may have role in reducing heavy metal toxicity by compartmentalization (Kim *et al.* 2013). *Beauveria caledonica*, a fungus that over-produces oxalic acid, has been shown to solubilize Cd, Cu, Zn and Pb via organic acid chelation. The mobilized metals were precipitated by the fungus into metal oxalate crystals that accumulated extracellularly and within the fungal biomass (Fomina *et al.* 2005).

Moreover, the drop in pH caused by the microbial secretion of organic acids, which occurs when key nutrients like phosphorus (P) are low in availability, can mobilize heavy metals adsorbed onto the soil matrix, increasing their bio-availability (Karimzadeh *et al.* 2012). P-solubilizing rhizosphere microbes have been shown to aid the mobilization of Zn and

Cd in Tris-minimal salts media by altering the pH of the medium (Li *et al.* 2010). Concentrations of labile Zn and Cd increased up to 86% and 88% respectively following a drop in pH (pH 7 down to 3) which was accompanied by the production of several organic acids including oxalic acid, tartaric acid, formic acid and acetic acid (Li *et al.* 2010). Within a pH range of 4-4.5, a drop of just 0.2 pH units can increase Cd solubility up to 5-fold (Kabata-Pendias 2011).

Siderophore Production

Microorganisms produce low molecular-weight (500-1000 daltons) complexes called siderophores in order to solubilize and sequester iron (Fe). Indeed, despite being abundant in the Earth's crust, iron is predominantly found in its ferric form (Fe^{3+}), which is sparingly bioavailable at biological pH (Neilands 1995, Glick 2003, Ma *et al.* 2011). Whilst plants produce their own phytochelatins to improve Fe^{3+} uptake, they can also utilize microbial siderophores, which have a much higher affinity for Fe^{3+} (Burd *et al.* 1998, Glick 2003). Given the ability of siderophores to solubilize and increase the bioavailability of metal cations other than Fe^{3+} (Neilands 1981, Hernlem *et al.* 1996, Rajkumar *et al.* 2010, Ma *et al.* 2011), they can also be used to improve phytoextraction. Indeed, the addition of the microbial siderophore, desferrioxamine-B (DFO-B) was able to improve the accumulation of Cd in the plant *Noccaea caerulescens* (formerly *Thlaspi*) by 37% (Karimzadeh *et al.* 2012). Moreover, the addition of active rhizosphere communities to *N. caerulescens* affected a 4-fold increase in net Zn hyper-accumulation due to the microbial mobilization of non-labile Zn pools. Following a lack of evidence to support mobilization via acid production, it was concluded that siderophores were most likely responsible for the increase in labile Zn (Whiting *et al.* 2001).

Biosurfactants

Biological surface-active agents, biosurfactants, are amphiphilic molecules that can bind to and alter the electrochemical properties of a surface. Microorganisms, including yeast species, produce a wide variety of biosurfactants that can increase the solubility of many contaminants (Ron and Rosenberg 2001). The hydrophobic proportion of the biosurfactant interacts with insoluble contaminants whilst the hydrophilic head group facilitates interactions with the bulk medium, forming a soluble surfactant-contaminant complex (Mulligan 2005).

Rhamolipid, a well-studied low-molecular-weight biosurfactant produced by *Pseudomonas aeruginosa* ATCC 9027, has been shown to assist in the removal of heavy metals and enhance the accumulation of Cu and Cd in rye grass (*Lolium perenne*)(Mulligan 2005, Gunawardana *et al.* 2011). Tea saponin, a seed-derived biosurfactant, has likewise been shown to improve the uptake of heavy metals, specifically of Cd, into corn (*Zea mays*) in a concentration-dependent manner (Xia *et al.* 2012). The positive impact of biosurfactants on phytoextraction may be manifold, as some biosurfactants also exhibit antimicrobial activity and can indirectly promote plant growth by inhibiting the growth of pathogens in the rhizosphere (Nielsen and Sørensen 2003, Lugtenberg and Kamilova 2009). However, as with all the strategies described for increasing the bioavailability of heavy metals, one must also be aware that there is an associated risk that mobilized metals will be leached into groundwater if the rate of plant uptake is not matched to the rate of mobilization (Angle and Linacre 2005, Bolan *et al.* 2014).

MICROBIAL-PLANT FEEDBACK LOOPS IN BIORE-MEDIATION

Promoting plant growth is an alternative strategy for enhancing phytoextraction. A review of 103 microbial-influenced phytoextraction studies found that in 70% of cases microorganisms had the effect of promoting plant growth, whilst only in 35% of cases did microorganisms increase the concentration of metal in plant tissues (Wood *et al.* 2016). The figures suggest that the ability of microorganisms to influence phytoextraction by promoting plant growth are far more common that metal-mobilization. Processes that stimulate plant growth can improve phytoextraction by increasing the amount of harvestable, heavy-metal-containing biomass as well as the available below-ground surface area at which reactions, including heavy metal mobilization and uptake, can take place (Wood *et al.* 2016). For example, increases in plant-root biomass, following fertilizer application, have been associated with the improved accumulation of Ni in *Alyssum bertolonii* by as much as 300 % (Robinson *et al.* 1997). Similarly, inoculations of *Pseudomonas* fluorescens and

P. tolaasii have been shown to improve Cd-phytoextraction in canola by 72% and 107%, respectively by promoting plant growth, as Cd concentrations in the plant tissues remained constant (Dell'Amico *et al.* 2008). A number of microbial processes are known to be able to improve plant growth either directly or indirectly and could be harnessed to optimize phytoextraction.

ACC Deaminase Production

1-aminocyclopropane-1-carboxylic acid (ACC) is a precursor to the plant stress signaling hormone, ethylene (Glick 2003, Onofre-Lemus *et al.* 2009). The microbially-mediated catabolism of ACC to α-ketobutyrate and ammonia, results in the removal of ACC from the rhizosphere and in the creation of a chemical gradient that perpetuates further ACC efflux. This process results into a decreased concentration of ACC in plant tissues together with the inhibition of ethylene production and normal plant stress responses (Burd *et al.* 1998, Ma *et al.* 2009, Onofre-Lemus *et al.* 2009). ACC-deaminating bacteria can alleviate heavy metal stress in plants and have been shown to increase plant biomass in the presence of Ni without reducing heavy metal concentrations in the plant tissues (Burd *et al.* 1998).

Enhanced Nutrient Supply

Nitrogen (N) and phosphorus (P) are limiting resources for plant growth in many environments (Hayat *et al.* 2010, Park *et al.* 2011). Both the fixation of atmospheric N_2 by *Rhizobia* spp. in root nodules and non-symbiotic diazotrophs, and the microbial solubilization of organically bound P (via phosphatases) or inorganically bound P (via organic acids) can significantly improve plant growth (Boddey and Victoria 1986, Peoples and Craswell 1992, Jones 1998, Shoebitz *et al.* 2009). Pot trials have demonstrated that a heavy-metal-resistant N-fixing *Bradyrhizobium* sp. can promote the growth of greengram, increasing dry weight in soils containing 9780 mg Zn kg^{-1} or 580 mg Ni kg^{-1}. Although bio-inoculant decreased the concentration of Zn and Ni in plant shoots by 11% and 19% respectively, the increase in biomass created a net positive contribution to the total amount of heavy metal extracted per plant (Wani *et al.* 2007). The microbial mediated increase in phyto-available P has been shown to improve heavy metal uptake (Li *et al.* 2010, Usha Rani *et al.*

2011). Inoculation of the Cd hyper-accumulator, *Brassica juncae*, with P-solubilizing bacteria induced a 4.5-fold increase in plant biomass over an 8-week period and a 250% increase in Cd accumulation (Jeong *et al.* 2013).

Plant Hormones

Microbes produce and secrete an array of compounds that regulate plant growth including auxins, cytokinins and gibberellins (García de Salamone *et al.* 2001, Badri *et al.* 2009, Usha Rani *et al.* 2011). Auxin (indole-3-acetic acid, IAA), is the major plant hormone that controls apical dominance, phototropism, shoot elongation, lateral and adventitious root growth (Elobeid *et al.* 2012). The production of IAA has been repeatedly cited in the literature as a major contributing factor to heavy metal hyper-accumulation (Ma *et al.* 2009, Glick 2010, Ma *et al.* 2011). In plants, IAA expression is negatively regulated by the presence of heavy metals such as Cd, but microbial secretion of IAA can counteract the inhibitory effects of heavy metals, enabling sustained plant growth (Elobeid *et al.* 2012).

Gibberellins, which contribute to germination and carbohydrate redistribution, have likewise been proposed to counteract heavy metal toxicity (Moya *et al.* 1995). Studies using soybean inoculated with endophytic *Penicillum funiculosum* suggest that the microbial-derived gibberellins not only promote plant growth by modulating gene expression, but alleviate heavy metal toxicity by forming stable complexes with heavy metals which are adsorbed onto the fungi (Khan and Lee 2013). The exogenous application of cytokinins, which can delay senescence, mediate cell growth, leaf expansion and root to shoot communication, have been shown to increase the biomass of sunflowers (*Helianthus annuus*) in the presence of Hg, as well as Hg accumulation. However, cytokinin treatments were unable to significantly increase translocation of Hg from root to shoot (García de Salamone *et al.* 2001, Cassina *et al.* 2012).

CONCLUSIONS

The ability to reduce soil pollution is a modern global challenge, and to meet this demand, new innovative ideas and methods are required. In plants, together with their microbiomes, lies huge unexploited potential for purifying sites contaminated with both hydrocarbons and heavy metals, with phytoremediation and bioremediation having already received significant scientific and commercial attention as a cheap and effective alternative to clean up contaminated areas. Therefore, we predict that future studies aimed at deciphering the structure and composition of these vital associations will aid in quantifying the role of microorganisms in facilitating or constraining the remediation process, allowing for the design of environmental-friendly strategies for contaminants remediation.

SELECTED REFERENCES

1. Agency for Toxic Substances and Disease Registry (2015). Priority List of Hazardous Substances. Atlanta, Agency for Toxic Substances and Disease Registry.
2. Aislabie, J. and Deslippe, J. R. (2013). Soil microbes and their contribution to soil services. Ecosystem services in New Zealand: conditions and trends. J. R. Dymond. Lincoln, NZ, Manaaki Whenua Press, Landcare Research.
3. Anderson, R. T., Vrionis, H. A., Ortiz-Bernad, I., Resch, C. T., Long, P. E., Dayvault, R., . . . Lovley, D. R. (2003). "Stimulating the In Situ Activity of Geobacter Species to Remove Uranium from the Groundwater of a Uranium-Contaminated Aquifer." Applied and Environmental Microbiology 69(10): 5884-5891.
4. Angle, J. S. and Linacre, N. A. (2005). "Metal phytoextraction - A survey of potential risks." International Journal of Phytoremediation 7(3): 241-254.
5. Badri, D. V., Weir, T. L., van der Lelie, D. and Vivanco, J. M. (2009). "Rhizosphere chemical dialogues: plant-microbe interactions." Current Opinion in Biotechnology 20(6): 642-650.
6. Boddey, R. M. and Victoria, R. L. (1986). "Estimation of biological nitrogen fixation associated with Brachiaria and Paspalum grasses using15N labelled organic matter and fertilizer." Plant and Soil 90(1-3): 265-292.
7. Bolan, N., Kunhikrishnan, A. and Gibbs, J. (2013). "Rhizoreduction

of arsenate and chromate in Australian native grass, shrub and tree vegetation." Plant and Soil 367(1-2): 615-625.

8. Bolan, N., Kunhikrishnan, A., Thangarajan, R., Kumpiene, J., Park, J., Makino, T., . . . Scheckel, K. (2014). "Remediation of heavy metal(loid)s contaminated soils - To mobilize or to immobilize?" Journal of Hazardous Materials 266: 141-166.
9. Burd, G. I., Dixon, D. G. and Glick, B. R. (1998). "A plant growth-promoting bacterium that decreases nickel toxicity in seedlings." Applied and Environmental Microbiology 64(10): 3663-3668.
10. Campbell, C. D., Grayston, S. J. and Hirst, D. J. (1997). "Use of rhizosphere carbon sources in sole carbon source tests to discriminate soil microbial communities." Journal of Microbiological Methods 30(1): 33-41.
11. Carmona, M., Zamarro, M. T., Biázquez, B., Durante-Rodríguez, G., Juárez, J. F., Valderrama, J. A., . . . Díaz, E. (2009). "Anaerobic catabolism of aromatic compounds: A genetic and genomic view." Microbiology and Molecular Biology Reviews 73(1): 71-133.
12. Cassina, L., Tassi, E., Pedron, F., Petruzzelli, G., Ambrosini, P. and Barbafieri, M. (2012). "Using a plant hormone and a thioligand to improve phytoremediation of Hg-contaminated soil from a petrochemical plant." Journal of Hazardous Materials 231-232: 36-42.
13. Chebbi, A., Hentati, D., Zaghden, H., Baccar, N., Rezgui, F., Chalbi, M., . . . Chamkha, M. (2017). "Polycyclic aromatic hydrocarbon degradation and biosurfactant production by a newly isolated Pseudomonas sp. strain from used motor oil-contaminated soil." International Biodeterioration and Biodegradation 122: 128-140.
14. Chen, B., Shen, J., Zhang, X., Pan, F., Yang, X. and Feng, Y. (2014). "The endophytic bacterium, sphingomonas samr12, improves the potential for zinc phytoremediation by its host, sedum alfredii." PLoS ONE 9(9).
15. Dell'Amico, E., Cavalca, L. and Andreoni, V. (2008). "Improvement of Brassica napus growth under cadmium stress by cadmium-resistant rhizobacteria." Soil Biology and Biochemistry 40(1): 74-84.
16. Ederer, M. M., Crawford, R. L., Herwig, R. P. and Orser, C. S.

(1997). "PCP degradation is mediated by closely related strains of the genus Sphingomonas." Molecular Ecology 6(1): 39-49.

17. Eggen, T. (1999). "Application of fungal substrate from commercial mushroom production - Pleuorotus ostreatus - For bioremediation of creosote contaminated soil." International Biodeterioration and Biodegradation 44(2-3): 117-126.
18. Elobeid, M., Göbel, C., Feussner, I. and Polle, A. (2012). "Cadmium interferes with auxin physiology and lignification in poplar." Journal of Experimental Botany 63(3): 1413-1421.
19. Floudas, D., Binder, M., Riley, R., Barry, K., Blanchette, R. A., Henrissat, B., . . . Hibbett, D. S. (2012). "The paleozoic origin of enzymatic lignin decomposition reconstructed from 31 fungal genomes." Science 336(6089): 1715-1719.
20. Fomina, M., Hillier, S., Charnock, J. M., Melville, K., Alexander, I. J. and Gadd, G. M. (2005). "Role of oxalic acid overexcretion in transformations of toxic metal minerals by Beauveria caledonica." Applied and Environmental Microbiology 71(1): 371-381.
21. Gadd, G. M. (2004). "Microbial influence on metal mobility and application for bioremediation." Geoderma 122(2-4 SPEC. IIS.): 109-119.
22. García de Salamone, I. E., Hynes, R. K. and Nelson, L. M. (2001). "Cytokinin production by plant growth promoting rhizobacteria and selected mutants." Canadian Journal of Microbiology 47(5): 404-411.
23. Gerhardt, K. E., Huang, X. D., Glick, B. R. and Greenberg, B. M. (2009). "Phytoremediation and rhizoremediation of organic soil contaminants: Potential and challenges." Plant Science 176(1): 20-30.
24. Glick, B. R. (2003). "Phytoremediation: Synergistic use of plants and bacteria to clean up the environment." Biotechnology Advances 21(5): 383-393.
25. Glick, B. R. (2010). "Using soil bacteria to facilitate phytoremediation." Biotechnology Advances 28(3): 367-374.
26. Gunawardana, B., Singhal, N. and Johnson, A. (2011). "Effects of amendments on copper, cadmium, and lead phytoextraction by Lolium perenne from multiple-metal contaminated solution."

International Journal of Phytoremediation 13(3): 215-232.

27. Guo, G., Tian, F., Ding, K., Wang, L., Liu, T. and Yang, F. (2017). "Effect of a bacterial consortium on the degradation of polycyclic aromatic hydrocarbons and bacterial community composition in Chinese soils." International Biodeterioration and Biodegradation 123: 56-62.
28. Halden, R. U., Halden, B. G. and Dwyer, D. F. (1999). "Removal of dibenzofuran, dibenzo-p-dioxin, and 2-chlorodibenzo-p-dioxin from soils inoculated with Sphingomonas sp. strain RW1." Applied and Environmental Microbiology 65(5): 2246-2249.
29. Hamamura, N., Olson, S. H., Ward, D. M. and Inskeep, W. P. (2006). "Microbial population dynamics associated with crude-oil biodegradation in diverse soils." Applied and Environmental Microbiology 72(9): 6316-6324.
30. Haritash, A. K. and Kaushik, C. P. (2009). "Biodegradation aspects of Polycyclic Aromatic Hydrocarbons (PAHs): A review." Journal of Hazardous Materials 169(1-3): 1-15.
31. Hayat, R., Ali, S., Amara, U., Khalid, R. and Ahmed, I. (2010). "Soil beneficial bacteria and their role in plant growth promotion: A review." Annals of Microbiology 60(4): 579-598.
32. Hernlem, B. J., Vane, L. M. and Sayles, G. D. (1996). "Stability constants for complexes of the siderophore desferrioxamine b with selected heavy metal cations." Inorganica Chimica Acta 244(2): 179-184.
33. Järup, L. (2003). "Hazards of heavy metal contamination." British Medical Bulletin 68: 167-182.
34. Jeong, S., Moon, H. S., Shin, D. and Nam, K. (2013). "Survival of introduced phosphate-solubilizing bacteria (PSB) and their impact on microbial community structure during the phytoextraction of Cd-contaminated soil." Journal of Hazardous Materials 263: 441-449.
35. Johnsen, A. R., Wick, L. Y. and Harms, H. (2005). "Principles of microbial PAH-degradation in soil." Environmental Pollution 133(1): 71-84.
36. Jones, D. L. (1998). "Organic acids in the rhizosphere - A critical review." Plant and Soil 205(1): 25-44.

37. Kabata-Pendias, A. (2011). Trace Elements in Soils and Plants. Florida, CRC Press.

38. Karimzadeh, L., Heilmeier, H. and Merkel, B. J. (2012). "Effect of microbial siderophore DFO-B on Cd accumulation by Thlaspi caerulescens hyperaccumulator in the presence of zeolite." Chemosphere 88(6): 683-687.

39. Kashefi, K. and Lovley, D. R. (2000). "Reduction of Fe(III), Mn(IV), and toxic metals at 100°C by Pyrobaculum islandicum." Applied and Environmental Microbiology 66(3): 1050-1056.

40. Kästner, M., Breuer-Jammali, M. and Mahro, B. (1994). "Enumeration and characterization of the soil microflora from hydrocarbon-contaminated soil sites able to mineralize polycyclic aromatic hydrocarbons (PAH)." Applied Microbiology and Biotechnology 41(2): 267-273.

41. Kazunga, C. and Aitken, M. D. (2000). "Products from the incomplete metabolism of pyrene by polycyclic aromatic hydrocarbon-degrading bacteria." Applied and Environmental Microbiology 66(5): 1917-1922.

42. Khan, A. L. and Lee, I. J. (2013). "Endophytic Penicillium funiculosum LHL06 secretes gibberellin that reprograms Glycine max L. growth during copper stress." BMC Plant Biology 13(1).

43. Khan, A. L., Waqas, M., Kang, S. M., Al-Harrasi, A., Hussain, J., Al-Rawahi, A., . . . Lee, I. J. (2014). "Bacterial endophyte Sphingomonas sp. LK11 produces gibberellins and IAA and promotes tomato plant growth." Journal of Microbiology 52(8): 689-695.

44. Kim, J. O., Lee, Y. W. and Chung, J. (2013). "The role of organic acids in the mobilization of heavy metals from soil." KSCE Journal of Civil Engineering 17(7): 1596-1602.

45. Kong, J., Wang, H., Liang, L., Li, L., Xiong, G. and Hu, Z. (2017). "Phenanthrene degradation by the bacterium Pseudomonas stutzeri JP1 under low oxygen condition." International Biodeterioration and Biodegradation 123: 121-126.

46. Kotterman, M. J. J., Vis, E. H. and Field, J. A. (1998). "Successive mineralization and detoxification of benzo[a]pyrene by the white rot fungus Bjerkandera sp. strain BOS55 and indigenous microflora." Applied and Environmental Microbiology 64(8): 2853-2858.

47. Li, W. C., Ye, Z. H. and Wong, M. H. (2010). "Metal mobilization and production of short-chain organic acids by rhizosphere bacteria associated with a Cd/Zn hyperaccumulating plant, Sedum alfredii." Plant and Soil 326(1): 453-467.

48. Lin, C., Owen, S. M. and Peñuelas, J. (2007). "Volatile organic compounds in the roots and rhizosphere of Pinus spp." Soil Biology and Biochemistry 39(4): 951-960.

49. Liu, H., Probst, A. and Liao, B. (2005). "Metal contamination of soils and crops affected by the Chenzhou lead/zinc mine spill (Hunan, China)." Science of the Total Environment 339(1-3): 153-166.

50. Liu, W., Luo, Y., Teng, Y., Li, Z. and Ma, L. Q. (2010). "Bioremediation of oily sludge-contaminated soil by stimulating indigenous microbes." Environmental geochemistry and health 32(1): 23-29.

51. Lovley, D. R., Holmes, D. E. and Nevin, K. P. (2004). "Dissimilatory Fe(III) and Mn(IV) reduction." Advances in Microbial Physiology 49: 219-286.

52. Lugtenberg, B. and Kamilova, F. (2009). Plant-growth-promoting rhizobacteria. 63: 541-556.

53. Lum, A. F., Ngwa, E. S. A., Chikoye, D. and Suh, C. E. (2014). "Phytoremediation Potential of Weeds in Heavy Metal Contaminated Soils of the Bassa Industrial Zone of Douala, Cameroon." International Journal of Phytoremediation 16(3): 302-319.

54. Ma, Y., Prasad, M. N. V., Rajkumar, M. and Freitas, H. (2011). "Plant growth promoting rhizobacteria and endophytes accelerate phytoremediation of metalliferous soils." Biotechnology Advances 29(2): 248-258.

55. Ma, Y., Rajkumar, M. and Freitas, H. (2009). "Improvement of plant growth and nickel uptake by nickel resistant-plant-growth promoting bacteria." Journal of Hazardous Materials 166(2-3): 1154-1161.

56. McLaughlin, M. J., Parker, D. R. and Clarke, J. M. (1999). "Metals and micronutrients - Food safety issues." Field Crops Research 60(1-2): 143-163.

57. Morita, A., Yanagisawa, O., Maeda, S., Takatsu, S. and Ikka, T. (2011). "Tea plant (Camellia sinensis L.) roots secrete oxalic acid

and caffeine into medium containing aluminum." Soil Science and Plant Nutrition 57(6): 796-802.

58. Moya, J. L., Ros, R. and Picazo, I. (1995). "Heavy metal-hormone interactions in rice plants: Effects on growth, net photosynthesis, and carbohydrate distribution." Journal of Plant Growth Regulation 14(2): 61-67.
59. Mrozik, A., Piotrowska-Seget, Z. and Łabuzek, S. (2003). "Bacterial degradation and bioremediation of polycyclic aromatic hydrocarbons." Polish Journal of Environmental Studies 12(1): 15-25.
60. Muehe, E. M., Obst, M., Hitchcock, A., Tyliszczak, T., Behrens, S., Schröder, C., . . . Kappler, A. (2013). "Fate of Cd during microbial Fe(III) mineral reduction by a novel and Cd-tolerant geobacter species." Environmental Science and Technology 47(24): 14099-14109.
61. Mulligan, C. N. (2005). "Environmental applications for biosurfactants." Environmental Pollution 133(2): 183-198.
62. Nagajyoti, P. C., Lee, K. D. and Sreekanth, T. V. M. (2010). "Heavy metals, occurrence and toxicity for plants: A review." Environmental Chemistry Letters 8(3): 199-216.
63. Neilands, J. B. (1981). "Microbial iron compounds." Annual Review of Biochemistry 50: 715-731.
64. Neilands, J. B. (1995). "Siderophores: Structure and function of microbial iron transport compounds." Journal of Biological Chemistry 270(45): 26723-26726.
65. Nielsen, T. H. and Sørensen, J. (2003). "Production of cyclic lipopeptides by Pseudomonas fluorescens strains in bulk soil and in the sugar beet rhizosphere." Applied and Environmental Microbiology 69(2): 861-868.
66. Novaes, E., Kirst, M., Chiang, V., Winter-Sederoff, H. and Sederoff, R. (2010). "Lignin and biomass: A negative correlation for wood formation and lignin content in trees." Plant Physiology 154(2): 555-561.
67. Ohtsuka, T., Yamaguchi, N., Makino, T., Sakurai, K., Kimura, K., Kudo, K., . . . Amachi, S. (2013). "Arsenic dissolution from Japanese paddy soil by a dissimilatory arsenate-reducing bacterium geobacter

sp. OR-1." Environmental Science and Technology 47(12): 6263-6271.

68. Onofre-Lemus, J., Hernández-Lucas, I., Girard, L. and Caballero-Mellado, J. (2009). "ACC (1-aminocyclopropane-1-carboxylate) deaminase activity, a widespread trait in Burkholderia species, and its growth-promoting effect on tomato plants." Applied and Environmental Microbiology 75(20): 6581-6590.

69. Owen, S. M., Clark, S., Pompe, M. and Semple, K. T. (2007). "Biogenic volatile organic compounds as potential carbon sources for microbial communities in soil from the rhizosphere of Populus tremula." FEMS Microbiology Letters 268(1): 34-39.

70. Pan, F., Meng, Q., Wang, Q., Luo, S., Chen, B., Khan, K. Y., . . . Feng, Y. (2016). "Endophytic bacterium Sphingomonas SaMR12 promotes cadmium accumulation by increasing glutathione biosynthesis in Sedum alfredii Hance." Chemosphere 154: 358-366.

71. Parales, R. E. and Harwood, C. S. (2002). "Bacterial chemotaxis to pollutants and plant-derived aromatic molecules." Current Opinion in Microbiology 5(3): 266-273.

72. Park, J. H. and Bolan, N. (2013). "Lead immobilization and bioavailability in microbial and root interface." Journal of Hazardous Materials 261: 777-783.

73. Park, J. H., Bolan, N., Megharaj, M. and Naidu, R. (2011). "Isolation of phosphate solubilizing bacteria and their potential for lead immobilization in soil." Journal of Hazardous Materials 185(2-3): 829-836.

74. Peoples, M. B. and Craswell, E. T. (1992). "Biological nitrogen fixation: Investments, expectations and actual contributions to agriculture." Plant and Soil 141(1-2): 13-39.

75. Rajkumar, M., Ae, N., Prasad, M. N. V. and Freitas, H. (2010). "Potential of siderophore-producing bacteria for improving heavy metal phytoextraction." Trends in Biotechnology 28(3): 142-149.

76. Rajkumar, M., Prasad, M. N. V., Freitas, H. and Ae, N. (2009). "Biotechnological applications of serpentine soil bacteria for phytoremediation of trace metals." Critical Reviews in Biotechnology 29(2): 120-130.

77. Robinson, B. H., Chiarucci, A., Brooks, R. R., Petit, D., Kirkman,

J. H., Gregg, P. E. H. and De Dominicis, V. (1997). "The nickel hyperaccumulator plant Alyssum bertolonii as a potential agent for phytoremediation and phytomining of nickel." Journal of Geochemical Exploration 59(2): 75-86.

78. Ron, E. Z. and Rosenberg, E. (2001). "Natural roles of biosurfactants." Environmental Microbiology 3(4): 229-236.

79. Salt, D. E., Smith, R. D. and Raskin, I. (1998). "Phytoremediation." Annual Review of Plant Physiology and Plant Molecular Biology 49: 643-668.

80. Schneider, J., Grosser, R., Jayasimhulu, K., Xue, W. and Warshawsky, D. (1996). "Degradation of pyrene, benz[a]anthracene, and benzo[a]pyrene by Mycobacterium sp. strain RJGII-135, isolated from a former coal gasification site." Applied and Environmental Microbiology 62(1): 13-19.

81. Shoebitz, M., Ribaudo, C. M., Pardo, M. A., Cantore, M. L., Ciampi, L. and Curá, J. A. (2009). "Plant growth promoting properties of a strain of Enterobacter ludwigii isolated from Lolium perenne rhizosphere." Soil Biology and Biochemistry 41(9): 1768-1774.

82. Siciliano, S. D., Fortin, N., Mihoc, A., Wisse, G., Labelle, S., Beaumier, D., . . . Greer, C. W. (2001). "Selection of Specific Endophytic Bacterial Genotypes by Plants in Response to Soil Contamination." Applied and Environmental Microbiology 67(6): 2469-2475.

83. Singh, O. V., Labana, S., Pandey, G., Budhiraja, R. and Jain, R. K. (2003). "Phytoremediation: An overview of metallic ion decontamination from soil." Applied Microbiology and Biotechnology 61(5-6): 405-412.

84. Stolz, J. F. and Oremland, R. S. (1999). "Bacterial respiration of arsenic and selenium." FEMS Microbiology Reviews 23(5): 615-627.

85. Sugiyama, A., Sano, C. M., Yazaki, K. and Sano, H. (2016). "Caffeine fostering of mycoparasitic fungi against phytopathogens." Plant Signaling and Behavior 11(1).

86. Sukweenadhi, J., Kim, Y. J., Kang, C. H., Farh, M. E. A., Nguyen, N. L., Hoang, V. A., . . . Yang, D. C. (2015). "Sphingomonas panaciterrae sp. nov. a plant growth-promoting bacterium isolated

from soil of a ginseng field." Archives of Microbiology 197(8): 973-981.

87. Topolska, J., Latowski, D., Kaschabek, S., Manecki, M., Merkel, B. J. and Rakovan, J. (2013). "Pb remobilization by bacterially mediated dissolution of pyromorphite Pb5(PO4)3Cl in presence of phosphate-solubilizing Pseudomonas putida." Environmental Science and Pollution Research: 1-11.

88. Usha Rani, M., Arundhathi and Reddy, G. (2011). "Bacillus cereus and Enterobacter cancerogenus screened for their efficient plant growth promoting traits rhizobacteria (PGPR) and antagonistic traits among sixteen bacterial isolates from rhizospheric soils of Pigeon Pea." African Journal of Microbiology Research 5(15): 2090-2094.

89. Vamerali, T., Bandiera, M. and Mosca, G. (2010). "Field crops for phytoremediation of metal-contaminated land. A review." Environmental Chemistry Letters 8(1): 1-17.

90. Walter, U., Beyer, M., Klein, J. and Rehm, H. J. (1991). "Degradation of pyrene by Rhodococcus sp. UW1." Applied Microbiology and Biotechnology 34(5): 671-676.

91. Wani, P. A., Khan, M. S. and Zaidi, A. (2007). "Effect of metal tolerant plant growth promoting Bradyrhizobium sp. (vigna) on growth, symbiosis, seed yield and metal uptake by greengram plants." Chemosphere 70(1): 36-45.

92. Wasi, S., Tabrez, S. and Ahmad, M. (2013). "Toxicological effects of major environmental pollutants: An overview." Environmental Monitoring and Assessment 185(3): 2585-2593.

93. Whiting, S. N., De Souza, M. P. and Terry, N. (2001). "Rhizosphere bacteria mobilize Zn for hyperaccumulation by Thlaspi caerulescens." Environmental Science and Technology 35(15): 3144-3150.

94. Wood, J. L., Liu, W., Tang, C. and Franks, A. E. (2016). "Microorganisms in heavy metal bioremediation: strategies for applying microbial-community engineering to remediate soils." AIMS Bioengineering 3(2): 211-229.

95. Wood, J. L., Tang, C. and Franks, A. E. (2016). "Microbial associated plant growth and heavy metal accumulation to improve phytoextraction of contaminated soils." Soil Biology and Biochemistry 103: 131-137.

96. Xia, H., Yan, Z. and Yu, B. (2012). Effect of tea saponin on the uptake of cadmium by corn from soil. 399-401: 1697-1701.

97. Zhuang, K., Ma, E., Lovley, D. R. and Mahadevan, R. (2012). "The design of long-term effective uranium bioremediation strategy using a community metabolic model." Biotechnology and Bioengineering 109(10): 2475-2483.

5

CHAPTER

MICROBES FOR ECOLOGICAL MONITORING, CONSERVATION AND RESTORATION

Jennifer Wood[1], Eleonora Egidi[2], Prashant Pant[3], Prerna Pandey[4], Pratibha Pant[5]

[1]Jennifer Wood is post-doctoral fellow in Dr. Ashley Franks' Applied and Environmental Microbiology Laboratory at La Trobe University, Australia. She has recently completed her Doctoral research in Microbial Ecology at the same university. Her broad research interest is in understanding the ecology of soil-microbial communities and their interactions with macro-communities. She is interested in developing trait-based approaches to interpret and predict how environmental and anthropogenic changes impact microbial community ecology and function.

[2] Dr. Eleonora Egidi is a mycologist interested in different aspects of fungal research, including fungal ecology, biology and phylogeny. Eleonora has an international background and has been involved in collaborative projects for the identification of new fungal barcoding regions, study of fungal communities from extreme environments, and use of fungi in phytoremediation. Her current scientific interests focus on exploring the biodiversity of Australian fungal flora and their role as drives of plant diversity in natural ecosystems, as well as understanding how these systems respond to environmental changes.

[3] Dr. Prashant Pant, is an Assistant Professor of Botany in Dyal Singh College, University of Delhi, New Delhi, India. He has been teaching for more than 5 years now and his main areas have been Bioinformatics, Molecular Biology and Genomics at the college. His PhD was on understanding the Microbial Succession using a Metagenomic approach in

a Barren mined-out site undergoing ecological restoration. For this, he deployed many molecular and computational Biology tool and techniques. He has several publications (articles and book chapters) in International Journals and International Publications. Currently, he is also Chief Editor of Bioinformatics Review and a co-founder/Executive Chief Editor of a Science Journal in Hindi Language "Swadeshi Vigyan" for the native language speaking communities. Both the Journals are a service to the scientific community as they are fully open access.

[4]Dr Prerna Pandey, a biotechnologist with several years of wet lab research experience, worked at International Center for Genetic Engineering and Biotechnology, New Delhi. Her field of research in PhD included isolation and molecular characterization of geminiviruses, genome sequencing, gene annotation, and gene silencing using the RNA interference technology. She has also worked at Transasia Biomedicals and Advance Enzyme Technologies as a scientist. Prerna has published papers in peer-reviewed journals, and has submitted a number of annotated Geminiviral genome sequences in the GenBank, including two novel ones. She has also completed her editing and proof-reading courses from Society for the promotion of Editors and Proof-readers, UK and now works as a freelance scientific editor and writer. When Prerna is not busy with her assignments, she loves to explore historical places.

[5]Dr. Pratibha Pant, is an Assistant Professor of Botany in Daulat Ram College, University of Delhi, New Delhi, India and has a teaching experience of more than 6 years. She is a plant virologist and biotechnologist by training and has done her PhD in Characterization of Begomoviral genomes from Papaya from Department of Botany, University of Delhi. During her PhD, she utilized a combination of molecular biology and Computational Biology tools to analyze begomoviral genomes and to her credit, a couple of novel begomoviruses have been isolated, sequenced and reported in public databases.

INTRODUCTION

Ecological restoration encompasses a spectrum of processes with the level of land degradation defining which set of processes and goals are most immediately important at a given site (Heneghan et al., 2008). In this framework, microbe-mediated processes are essential elements of ecological restoration. The establishment, growth, productivity and desired trajectories of succession of native plant communities at ecological restoration sites towards natural forested reference states requires re-building healthy, resilient soils and below ground biota. Complex networks that connect above ground and below ground ecosystems involve the rhizosphere in processes of mineralization and nutrient cycling. Throughout this book, we have highlighted the essential roles of microorganisms in soil function and in generating soil structure (chapter 2); the regulation of plant productivity and community dynamics (chapter 3); and the removal and transformation of soil contaminants (chapter 4). Yet it has only been in the last decade that soil microorganisms have emerged as potential tool to be used in restoration ecology (Harris, 2009). The way in which microorganisms are used in restoration can be classified into two broad categories: (1) microbial functions can be harnessed remove physical (i.e. soil structure), biotic (i.e. invasive plant species) or chemical (i.e. contaminants) barriers which impede restoration progress; (2) microorganisms can be used as bio-indicators of restoration success (Harris, 2009). In this chapter, we will discuss how the different roles of soils microorganisms (outlined in chapters 2-4) pertain to various goals in ecological restoration as well as the strategies by which microorganisms can be utilized in ecological restoration. Figure 5.1 illustrates a scheme for the inter-relationships of biotic communities during Ecological restoration. The below ground rhizospheric bacteria, fungi and micro-fauna respond to the above ground communities and soil types in which they occur. The interactions therein during eco-restoration likely involve utilization of plant exudates, carbon sequestration, decomposition of plant/animal litter and forging trophic levels within the soil food web. Shifts in the numbers and composition of rhizospheric biota of native plants can be used as markers for the trajectories of below-ground changes occurring at sites undergoing eco-restoration. Thus, rhizospheric soil biota is seen as hyper-diverse and dynamic with seasonal shifts and temporal changes marked by shift in above ground plant communities.

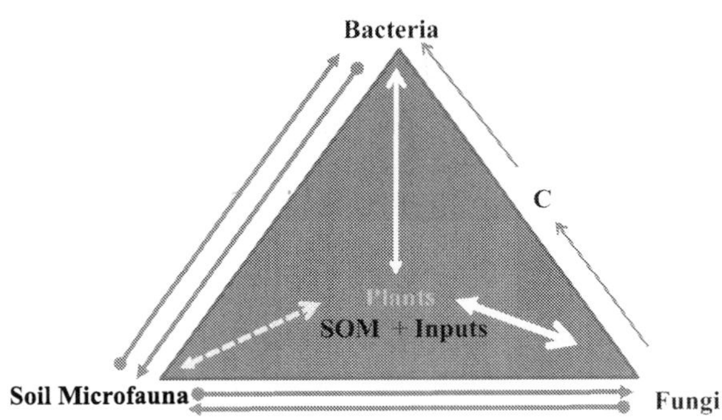

Figure 5.1: A scheme for the inter-relationships of biotic communities during ecological restoration.

PUTTING MICROORGANISMS TO WORK IN RESTORATION

The way in which microbial functions can be utilized in ecological restoration is best considered in the context of the severity of the original perturbation and the goal of the restoration project. The goals of ecological restoration range from reclamation through to rehabilitation. Rehabilitation refers to processes aimed at returning degraded land to the state of the original ecosystem prior to degradation (Heneghan et al., 2008). A similar, but contrasting goal in restoration is replacement, which refers to the restoration of degraded land to a functioning ecosystem, but not necessarily to the same ecosystem that was present prior to degradation. Reclamation is the process of returning land to some form of beneficial use. For instance, at highly degraded sites reclamation processes may simply be aimed at improving soil physical or chemical properties in order to facilitate some revegetation (Heneghan et al., 2008).

PUTTING MICROORGANISMS TO WORK DURING RECLAMATION

Some soils can become so degraded that plant establishment is not possible without an initial soil reclamation step. For example, in areas that are geologically fragile and over populated, intensive land-use can lead to a form of degradation that consists of serious soil erosion and eventual exposure of the bedrock known as rocky desertification. These landscapes have poor soil fertility (because there is little soil left) and plant establishment is impeded because there is no topsoil to retain the nutrients plants need to grow; the result is a desert-like landscape (Wu et al., 2017). Rocky desertification is a major issue in the rocky mountain region of Southwest China which mainly consists of carbonate bedrock.

Biological soil crusts (BSC) have been shown to assist in the formation of new soil from exposed carbonate bedrock. BSCs are diverse communities usually dominated by cyanobacteria, which also contain green and brown algae, mosses, lichens, liverworts, bacteria and fungi. The BSC can assist soil formation in two ways. Firstly, by capturing and incorporating dust and other fine particles which are blown onto the community (Chen et al., 2014). Secondly, the re-wetting and drying of BSCs can accelerate the direct weathering of the underlying bedrock via the expansion and contraction of mucilaginous and gelatinous living material. This process exerts a mechanical force on the bedrock assisting in its breakdown (Chen et al., 2014).

Individual microorganisms can also assist in soil formation. Recently a bacterium, known as *Bacillus megatherium* NL-7, has been isolated that can accelerate the weathering of carbonate rock and promote soil formation via the secretion of acetic acid (Wu et al., 2017). The process not only dissolves the carbonate bedrock but releases inorganic P and K into the newly forming topsoil improving soil fertility for plant establishment (Wu et al., 2017). Thus, in addition to physical and chemical weathering processes, microorganisms have a significant role to play in soil formation and the remediation of highly degraded land.

Mine tailings, are the waste material left over from processing ore. Mine tailings are low in organic matter needed for plant re-establishment and often high in residual toxic metal ions which likewise impede plant establishment (Tordoff et al., 2000). As such, ecological restoration projects dealing with mine tailings often need an initial reclamation step before re-vegetation can begin. The mixing of mine tailings with organic material such as biosolids or manure is a rapid strategy for creating a soil matrix for plant establishment, but does not mitigate the toxic properties of the tailings.

Microbial driven metal solubilization is a potential strategy that can be used to pretreat mine tailings, prior to mixing them with organics material, to remove metals and metal toxicity. In particular chemo-lithotrophic bacteria (which obtain their energy from the oxidation of reduced inorganic compounds such as sulphide, ammonia and hydrogen) have been used to remove metals from contaminated substrate (White et al., 1998; Gadd, 2010). The process is known as bioleaching and the most common leaching bacteria belong to the genus *Acidithiobacillus* such as *A. ferrooxidans* and *A. thiooxidans*. These microorganisms can convert insoluble metal sulphides (e.g. FeS_2, MoS_2,) into soluble metal ions and liberate acid-labile forms of metals via the production of sulphuric acid (Bosecker, 1997). Being acidophilic *Acidithiobacillus spp.* are able to survive and thrive during the bioleaching process (Bosecker, 1997). Indigenous Thiobacilli that were preadapted to red mud (the main waste product of aluminium extraction from bauxite) were able to solubilize 6265 mg of aluminium/L in the presence of 5% v/v red mud and 1 % w/v sulphur added as an energy source (Vachon et al., 1994).

When bioleaching processes (sulphate oxidation by acidophilic microorganisms) occur in nature, metals can be mobilized from waste ores and generate acid mine drainage, which is of serious environmental concern. As such, bioleaching is not carried out *in situ* but in bioreactors. Bioleaching by sulphur-oxidizing bacteria can be coupled to bio-precipitation of metals by sulphate-reducing bacteria to remove toxic metal ions from the soil and then from the resulting effluent in a two-step process (White et al., 1997; Figure 1).

An alternate strategy that aims to reclaim heavy metal contaminated substrate *in situ* is phytoextraction (described in chapter 4). To reclaim severely contaminated soils, phytoextraction processes can utilize specialized plants, called heavy metal hyper-accumulators (HMHs; Salt et al., 1998). Unlike non-HMH plants, HMHs are often endemic to soils containing high levels of heavy metals and so, unlike most plants, can established and grow on highly contaminated substrate. Indeed, some Zn HMHs can only grow on highly contaminated soils as they sequester Zn so effectively that growth on uncontaminated soils results in Zn deficiencies (Ozturk et al., 2003). HMHs can obtain concentrations of heavy metals in their above-ground tissues 100-1000-fold higher than what is typically found in non-HMHs and, as such, can also facilitate the recovery of commodity heavy metals in a process termed 'phytomining' (Anderson et al., 1999; Robinson et al., 1999; Alford et al., 2010; Rascio and Navari-Izzo, 2011; Sheoran et al., 2011).

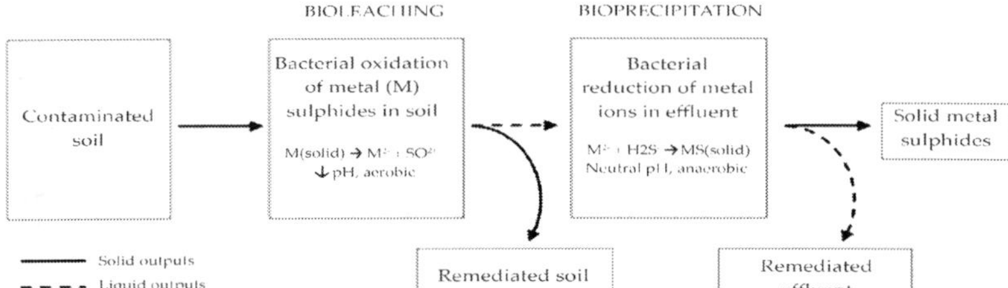

Figure 5.2: Diagram showing the integration of bacterial sulphate oxidation and bacterial sulphate reduction in the reclamation of heavy metal contaminated substrates, such as mine tailings. Arrows indicate inputs and outputs of

the bioleaching and bioprecipitation phases. Solid lines indicate solid inputs/outputs, dashed lines indicate liquid inputs/outputs. M = target metal. Adapted from (White et al., 1997).

The cooperative use of HMH plants and soil microorganisms to reclaim highly contaminated mine tailings is an attractive approach as it can be performed *in situ*, is low cost and low labour intensive, and is environmentally friendly compared to remediation strategies, such as bioleaching, which involve the physical removal of contaminated soil, washing and reburial. However, compared to bioleaching, phytoextraction can take years to remove sufficient levels of heavy metals for restoration to proceed. For this reason, the synergies between the HMH plant and soil microorganisms are being intensely studied in order to optimize the phytoextraction process. To date, multiple studies have provided evidence that rhizosphere-associated microorganisms have a positive impact on hyper-accumulation efficacy, increasing hyper-accumulation as much as 4-fold when compared to rates obtained in sterilized soil (De Souza et al., 1999; Whiting et al., 2001; Abou-Shanab et al., 2003a; Abou-Shanab et al., 2003b; Amprayn et al., 2012; Malekzadeh et al., 2012). A number of the processes by which they are hypothesised to do so are described in chapter 4.

PUTTING MICROORGANISMS TO WORK DURING REHABILITATION

The loss of soil physicochemical and biological properties reduces soil fertility which, when the degradation is severe, can culminate in the loss native vegetation cover (Requena et al., 2001). The loss of vegetation creates a positive feedback loop which accelerates the degradation process: the loss of plant cover accelerates the erosion and weathering of soil, which degrades soils further and reduces the potential for new plants to establish (Figure 2).

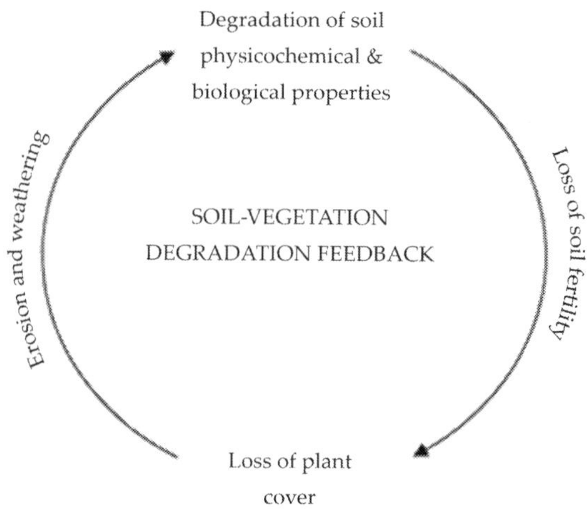

Figure 5.3: Positive soil-degradation feedback loop: degradation of soil physicochemical and biological properties reduces soil fertility which leads to a reduction in the ability of plants to establish. Loss of plant cover accelerates the erosion and weathering of soil, further degrading soils and further reducing the potential for new plants to establish.

Recent research using microcosm studies have shown that bacteria occupying the plant rhizosphere, which are likely partake in plant-microbe interactions, are adapted to compete for space and nutrients in the rhizosphere (Wood et al., 2017). From a microbiological perspective, soil degradation alters the soil environment to favor microorganisms that are adapted to survive in stress tolerant environments rather than competitive conditions and may result in the loss of plant-microbial partners from the bulk soil. The further loss of vegetation from degraded sites effectively removes from the environment the last reservoirs from which mutualistic microbial symbionts (which are crucial for governing plant nutrient cycles and sustaining plant cover) may be reintroduced. As such, numerous approaches in ecological restoration are emerging that aim to re-introduce soil microorganisms to promote the re-establishment of native plants and halt this cycle degradation (Requena et al., 2001). Historically, the most important known microbial symbionts have been mycorrhizal fungi and N-fixing rhizobia. Accordingly, these two groups of microorganisms have been the focus of microbial-assisted revegetation strategies in restoration ecology.

Mycorrhizal associations can assist plants to overcome physicochemical constraints that inhibit plant growth and establishment in numerous ways. Firstly, mycorrhizal associations can alleviate stress via biochemical interactions. For example, in saline soils arbuscular mycorrhizal symbioses have been shown to stimulate the accumulation of osmoregulators whilst in the presence of heavy metals mycorrhizal associations stimulated the induction of genes that encode for proteins thought to mitigate damage caused by reactive oxygen species, including heat shock proteins and glutathione *S*-transferase (Hildebrandt et al., 2007; Evelin et al., 2009). Secondly, the augmentation of plant-root systems by mycorrhizae can alleviate plant stress in saline and acid soils by counterbalancing the reduction in root surface area imposed by soil stressors. The result is improved plant nutritional uptake and, in turn, growth and establishment (Marschner, 1991). Thirdly, mycorrhizal association can even alter plant physiological responses to assist growth and establishment. For instance, in saline soils mycorrhizal associations improved plant photosynthetic rate and water-use efficiency (Evelin et al., 2009).

Mycorrhizae can be introduced into soil to aid ecological restoration and plant re-establishment via the addition of spores, the direct inoculation of plants to be introduced into the site, or the addition of soil inocula from undisturbed communities (Heneghan et al., 2008). As interest in utilizing mycorrhizal fungi in ecological restoration grows, important considerations are beginning to emerge regarding the effectiveness of commercial fungi and opposed to native fungi (Heneghan et al., 2008). Studies using mycorrhizae to aid re-vegetation of degraded semiarid ecosystems demonstrated that the source of mycorrhizal inoculum can significantly impact restoration success. Seedlings to be planted out into severely degraded habitat were inoculated with either an exotic AMF, *Glomus intraradices,* or a consortium of five mycorrhizae native to the restoration site (Requena et al., 2001). After five years, plants inoculated with native AMF were approximately double the size and height of plants inoculated with exotic AMF. Additionally, all native AMF propagules were present in the rhizosphere of 5-year old plants, while spores from introduced AMF were scarce, suggesting that the native inoculum was able to establish sustainable relationships with the transplants that would facilitate the maintenance of the inoculum within the ecosystem and assist future propagule establishment. Although they offer many benefits, it is important to note that the application of mycorrhizae during ecological restoration requires knowledge and consideration of the site conditions. For instance, mycorrhizae may not grow at sites contaminated with heavy metals and may be inhibited by high levels of nutrients such as nitrogen (Egerton-Warburton et al., 2007; Heneghan et al., 2008).

Symbiotic nitrogen-fixing bacteria such as *Rhizobium* spp. and *Bradyrhizobium* spp. have also been shown to alleviate plant stress and improve plant growth in the presence of salinity (Oufdou et al., 2014; Benidire et al., 2017), heavy metals (Wani et al., 2007; Sánchez-Pardo and Zornoza, 2014) and drought stress (Esfahani and Mostajeran, 2011). The impact of these bacteria upon plant growth is dependant on successful root infection which can be impacted by soil contaminants. Thus, as with mycorrhizae, the successful application of bacterial symbionts in ecological restoration requires knowledge about the restoration site and microbial physiology. Generally accepted it is accepted that bacteria are more susceptible to stressors such as heavy metals and soil pH compared to fungi (Hiroki, 1992; Rajapaksha et al., 2004; Rousk et al., 2010). As such, resistant strains of N-fixing bacteria may be necessary for successful restoration at severely degraded sites.

Inoculation of Acacia seeds with elite strains of native N-fixing bacterial symbionts has been shown to increase seedling establishment by 118 % (Thrall et al., 2005). In restoration projects which aim to re-establish native ecosystems, the sourcing of native seeds can be expensive and can impact negatively on the remaining intact ecosystems. Thus, these results not only indicate an important role for bacteria in restoration success, by in improving the cost efficiency and environmental sustainability of restoration projects.

MICROORGANISMS AS RESTORATION BIOINDICATORS

In addition to driving restoration processes, there has been a long history of using microorganisms as bio-indicators of restoration success. For instance, it is widely accepted that increased land use intensity (i.e. more degraded soils) causes a reduction in the fungal to bacterial (F:B) ratio, and that these ratios increase as natural ecosystems are restored

(Bardgett and McAlister, 1999; Harris, 2009). These observations are not always statistically significant (Lauber et al., 2008), but do suggest that fungi are less prevalent in managed or degraded ecosystems and higher fungal abundance is indicative of a healthy natural ecosystem. Interestingly the increase in F:B ratio with time does not appear to be linear. Using a chronosequence of 26 ex-arable soils ranging from 1–34 years of abandonment Van Der Wal et al.,. (2006) found that whilst fungal biomass increased in the first 2 years following abandonment, probably due to an initial increase in complex organic material entering the soil, no further increases in biomass were apparent even after a period of 30 years abandonment. This long term abandoned soils still exhibited lower values for fungal biomass, total and recalcitrant carbon compared to intact sites and the authors suggest that intervention was needed – specifically addressing abiotic soil properties – in order to further restore fungal biomass to levels observed in natural ecosystems (Van Der Wal et al., 2006).

Whilst F:B ratios are a broad metric for assessing ecosystem health, specific organisms have been suggested as bio-indicators for restoration success in various habitat types. In peatland restoration, testate amoebae have been suggested as a proxy bioindicator to assess water table depth and pH (Laggoun-Défarge et al., 2008). Testate amoebae are particularly useful because, in addition to providing information on the soil biota, they are preserved in the peat deposits, allowing for paleo-environmental reconstructions of peat communities and their analysis is inexpensive and relatively non-labour intensive (Laggoun-Défarge et al., 2008). As we learn more about the ecology of soil microorganisms, there is potential for specific fungal and bacterial species or functional traits to be used as indicators of remediation success.

CONCLUSIONS

Ecological restoration is a complex and multifaceted process. Each restoration project will be unique to the site to be restored and the restoration strategy will be informed by: the level and type of degradation, soil physico chemical properties, vegetation history and, of course, the goals, budget and time scale of the restoration project. Microorganisms have real and significant role to play in ecological restoration, whether it be reclaiming highly degraded soils, restoring native ecosystems or simply as a monitoring tool. Integrating microbial ecology into restoration projects will result in a deeper understanding and more holistic approach to ecological restoration that will no doubt increase restoration efficiency and success.

SELECTED REFERENCES

1. Abou-Shanab, R.A., Angle, J.S., Delorme, T.A., Chaney, R.L., Van Berkum, P., Moawad, H., et al.,. (2003a). Rhizobacterial effects on nickel extraction from soil and uptake by Alyssum murale. *New Phytologist* 158(1), 219-224.
2. Abou-Shanab, R.I., Delorme, T.A., Angle, J.S., Chaney, R.L., Ghanem, K., Moawad, H., et al.,. (2003b). Phenotypic Characterization of Microbes in the Rhizosphere of Alyssum murale. *International Journal of Phytoremediation* 5(4), 367-379.
3. Alford, É.R., Pilon-Smits, E.A.H., and Paschke, M.W. (2010). Metallophytes-a view from the rhizosphere. *Plant Soil.* 337(1), 33-50.
4. Amprayn, K.O., Rose, M.T., Kecskés, M., Pereg, L., Nguyen, H.T., and Kennedy, I.R. (2012). Plant growth promoting characteristics of soil yeast (Candida tropicalis HY) and its effectiveness for promoting rice growth. *Applied Soil Ecology* 61, 295-299.
5. Anderson, C.W.N., Brooks, R.R., Chiarucci, A., Lacoste, C.J., Leblanc, M., Robinson, B.H., et al.,. (1999). Phytomining for nickel, thallium and gold. *Journal of Geochemical Exploration* 67(1-3), 407-415.
6. Bardgett, R.D., and McAlister, E. (1999). The measurement of soil fungal:bacterial biomass ratios as an indicator of ecosystem self-regulation in temperate meadow grasslands. *Biology and Fertility of Soils* 29(3), 282-290. doi: 10.1007/s003740050554.
7. Benidire, L., Lahrouni, M., Khalloufi, F.E., Göttfert, M., and Oufdou, K. (2017). Effects of Rhizobium leguminosarum inoculation on growth, nitrogen uptake and mineral assimilation in Vicia faba plants under salinity stress. *Journal of Agricultural Science and Technology* 19(4), 889-901.

8. Bosecker, K. (1997). Bioleaching: metal solubilization by microorganisms. *FEMS Microbiol. Rev.* 20(3), 591-604. doi: http://dx.doi.org/10.1016/S0168-6445(97)00036-3.
9. Chen, Y., Lian, B., Yin, Z., and Tang, Y. (2014). Weathering of carbonate rocks by biological soil crusts in karst areas. *Journal of Earth Science* 25(4), 662-667. doi: 10.1007/s12583-014-0455-1.
10. De Souza, M.P., Huang, C.P.A., Chee, N., and Terry, N. (1999). Rhizosphere bacteria enhance the accumulation of selenium and mercury in wetland plants. *Planta* 209(2), 259-263.
11. Egerton-Warburton, L.M., Johnson, N.C., and Allen, E.B. (2007). Mycorrhizal community dynamics following nitrogen fertilization: A cross-site test in five grasslands. *Ecological Monographs* 77(4), 527-544. doi: 10.1890/06-1772.1.
12. Esfahani, M.N., and Mostajeran, A. (2011). Rhizobial strain involvement in symbiosis efficiency of chickpea-rhizobia under drought stress: Plant growth, nitrogen fixation and antioxidant enzyme activities. *Acta Physiologiae Plantarum* 33(4), 1075-1083. doi: 10.1007/s11738-010-0635-2.
13. Evelin, H., Kapoor, R., and Giri, B. (2009). Arbuscular mycorrhizal fungi in alleviation of salt stress: A review. *Annals of Botany* 104(7), 1263-1280. doi: 10.1093/aob/mcp251.
14. Gadd, G.M. (2010). Metals, minerals and microbes: Geomicrobiology and bioremediation. *Microbiology* 156(3), 609-643. doi: 10.1099/mic.0.037143-0.
15. Gerhardt, K.E., Huang, X.D., Glick, B.R., and Greenberg, B.M. (2009). Phytoremediation and rhizoremediation of organic soil contaminants: Potential and challenges. *Plant Science* 176(1), 20-30.
16. Harris, J. (2009). Soil microbial communities and restoration ecology: Facilitators or followers? *Science* 325(5940), 573-574. doi: 10.1126/science.1172975.
17. Heneghan, L., Miller, S.P., Baer, S., Callaham Jr, M.A., Montgomery, J., Pavao-Zuckerman, M., et al.,. (2008). Integrating soil ecological knowledge into restoration management. *Restoration Ecology* 16(4), 608-617. doi: 10.1111/j.1526-100X.2008.00477.x.
18. Hildebrandt, U., Regvar, M., and Bothe, H. (2007). Arbuscular

mycorrhiza and heavy metal tolerance. *Phytochemistry* 68(1), 139-146. doi: 10.1016/j.phytochem.2006.09.023.

19. Hiroki, M. (1992). Effects of heavy metal contamination on soil microbial population. *Soil Science and Plant Nutrition* 38(1), 141-147. doi: 10.1080/00380768.1992.10416961.
20. Laggoun-Défarge, F., Mitchell, E., Gilbert, D., Disnar, J.R., Comont, L., Warner, B.G., et al.,. (2008). Cut-over peatland regeneration assessment using organic matter and microbial indicators (bacteria and testate amoebae). *Journal of Applied Ecology* 45(2), 716-727. doi: 10.1111/j.1365-2664.2007.01436.x.
21. Lauber, C.L., Strickland, M.S., Bradford, M.A., and Fierer, N. (2008). The influence of soil properties on the structure of bacterial and fungal communities across land-use types. *Soil Biology and Biochemistry.* 40(9), 2407-2415. doi: 10.1016/j.soilbio.2008.05.021.
22. Liu, H., Probst, A., and Liao, B. (2005). Metal contamination of soils and crops affected by the Chenzhou lead/zinc mine spill (Hunan, China). *Science of the Total Environment* 339(1-3), 153-166.
23. Malekzadeh, E., Alikhani, H.A., Savaghebi-Firoozabadi, G.R., and Zarei, M. (2012). Bioremediation of cadmium-contaminated soil through cultivation of maize inoculated with plant growth-promoting rhizobacteria. *Bioremediation Journal.* 16(4), 204-211.
24. Marschner, H. (1991). Mechanisms of adaptation of plants to acid soils. *Plant. Soil.* 134(1), 1-20. doi: 10.1007/BF00010712.
25. Oufdou, K., Benidire, L., Lyubenova, L., Daoui, K., El Abidine Fatemi, Z., and Schröder, P. (2014). Enzymes of the glutathione-ascorbate cycle in leaves and roots of rhizobia-inoculated faba bean plants (Vicia faba L.) under salinity stress. *European Journal of Soil Biology* 60, 98-103. doi: 10.1016/j.ejsobi.2013.11.002.
26. Ozturk, L., Karanlik, S., Ozkutlu, F., Cakmak, I., and Kochian, L.V. (2003). Shoot biomass and zinc/cadmium uptake for hyperaccumulator and non-accumulator Thlaspi species in response to growth on a zinc-deficient calcareous soil. *Plant Science* 164(6), 1095-1101.
27. Rajapaksha, R.M.C.P., Tobor-Kapłon, M.A., and Bååth, E. (2004). Metal toxicity affects fungal and bacterial activities in soil differently. *Appl. Environ. Microbiol.* 70(5), 2966-2973.

28. Rascio, N., and Navari-Izzo, F. (2011). Heavy metal hyperaccumulating plants: How and why do they do it? And what makes them so interesting? *Plant Science* 180(2), 169-181.

29. Requena, N., Perez-Solis, E., Azcón-Aguilar, C., Jeffries, P., and Barea, J.M. (2001). Management of indigenous plant-microbe symbioses aids restoration of desertified ecosystems. *Applied Environmental Microbiology.* 67(2), 495-498. doi: 10.1128/AEM.67.2.495-498.2001.

30. Robinson, B.H., Brooks, R.R., Gregg, P.E.H., and Kirkman, J.H. (1999). The nickel phytoextraction potential of some ultramafic soils as determined by sequential extraction. *Geoderma* 87(3-4), 293-304.

31. Rousk, J., Bååth, E., Brookes, P.C., Lauber, C.L., Lozupone, C., Caporaso, J.G., et al.,. (2010). Soil bacterial and fungal communities across a pH gradient in an arable soil. *ISME J.* 4(10), 1340-1351. doi: 10.1038/ismej.2010.58.

32. Salt, D.E., Smith, R.D., and Raskin, I. (1998). "Phytoremediation".).

33. Sánchez-Pardo, B., and Zornoza, P. (2014). Mitigation of Cu stress by legume-Rhizobium symbiosis in white lupin and soybean plants. *Ecotoxicology and Environmental Safety* 102(1), 1-5. doi: 10.1016/j.ecoenv.2014.01.016.

34. Sheoran, V., Sheoran, A.S., and Poonia, P. (2011). Role of hyperaccumulators in phytoextraction of metals from contaminated mining sites: A review. *Critical Reviews in Environmental Science and Technology* 41(2), 168-214.

35. Thrall, P.H., Millsom, D.A., Jeavons, A.C., Waayers, M., Harvey, G.R., Bagnall, D.J., et al.,. (2005). Seed inoculation with effective root-nodule bacteria enhances revegetation success. *Journal of Applied Ecology* 42(4), 740-751. doi: 10.1111/j.1365-2664.2005.01058.x.

36. Tordoff, G.M., Baker, A.J.M., and Willis, A.J. (2000). Current approaches to the revegetation and reclamation of metalliferous mine wastes. *Chemosphere* 41(1), 219-228. doi: http://dx.doi.org/10.1016/S0045-6535(99)00414-2.

37. Vachon, P., Tyagi, R.D., Auclair, J.C., and Wilkinson, K.J. (1994). Chemical and biological leaching of aluminum from red mud.

Environmental Science & Technology 28(1), 26-30. doi: 10.1021/es00050a005.

38. Vamerali, T., Bandiera, M., and Mosca, G. (2010). Field crops for phytoremediation of metal-contaminated land. A review. *Environmental Chemistry Letters* 8(1), 1-17.
39. Van Der Wal, A., Van Veen, J.A., Smant, W., Boschker, H.T.S., Bloem, J., Kardol, P., et al.,. (2006). Fungal biomass development in a chronosequence of land abandonment. *Soil Biology and Biochemistry.* 38(1), 51-60. doi: 10.1016/j.soilbio.2005.04.017.
40. Wani, P.A., Khan, M.S., and Zaidi, A. (2007). Effect of metal tolerant plant growth promoting Bradyrhizobium sp. (vigna) on growth, symbiosis, seed yield and metal uptake by greengram plants. *Chemosphere* 70(1), 36-45.
41. White,C.,Sayer,J.A.,andGadd,G.M.(1997).Microbialsolubilization and immobilization of toxic metals: key biogeochemical processes for treatment of contamination. *FEMS Microbiol. Rev.* 20(3-4), 503-516. doi: 10.1111/j.1574-6976.1997.tb00333.x.
42. White, C., Shaman, A.K., and Gadd, G.M. (1998). An integrated microbial process for the bioremediation of soil contaminated with toxic metals. *Nature Biotechnolog* 16(6), 572-575.
43. Whiting, S.N., De Souza, M.P., and Terry, N. (2001). Rhizosphere bacteria mobilize Zn for hyperaccumulation by Thlaspi caerulescens. *Environmental Science and Technology* 35(15), 3144-3150.
44. Wood, J.L, Tang, C., Franks, A.E. (2017). Competitive traits are more important than stress tolerant traits in a cadmium-contaminated rhizosphere: a role for trait theory in microbial ecology. *Frontiers in Microbiology* (article in press)
45. Wu, Y.W., Zhang, J.C., Wang, L.J., and Wang, Y.X. (2017). A rock-weathering bacteria isolated from rock surface and its role in ecological restoration on exposed carbonate rocks. *Ecological Engineering* 101, 162-169. doi: 10.1016/j.ecoleng.2017.01.023.

6
CHAPTER

THE INTERACTIONS OF MICROBES WITH ALIEN, INVASIVE SPECIES

Prerna Pandey[1], Pratibha Pant[2], Prashant Pant[3]

[1]Dr Prerna Pandey, a biotechnologist with several years of wet lab research experience, worked at International Center for Genetic Engineering and Biotechnology, New Delhi. Her field of research in PhD included isolation and molecular characterization of geminiviruses, genome sequencing, gene annotation, and gene silencing using the RNA interference technology. She has also worked at Transasia Biomedicals and Advance Enzyme Technologies as a scientist. Prerna has published papers in peer-reviewed journals, and has submitted a number of annotated Geminiviral genome sequences in the GenBank, including two novel ones. She has also completed her editing and proof-reading courses from Society for the promotion of Editors and Proof-readers, UK and now works as a freelance scientific editor and writer. When Prerna is not busy with her assignments, she loves to explore historical places.

[2]Dr. Pratibha Pant, is an Assistant Professor of Botany in Daulat Ram College, University of Delhi, New Delhi, India and has a teaching experience of more than 6 years. She is a plant virologist and biotechnologist by training and has done her PhD in Characterization of Begomoviral genomes from Papaya from Department of Botany, University of Delhi. During her PhD, she utilized a combination of molecular biology and Computational Biology tools to analyze begomoviral genomes and to her credit, a couple of novel begomoviruses have been isolated, sequenced and reported in public databases.

[3] Dr. Prashant Pant, is an Assistant Professor of Botany in Dyal Singh College, University of Delhi, New Delhi, India. He has been teaching for more than 5 years now and his main areas have been Bioinformatics, Molecular Biology and Genomics at the college. His

PhD was on understanding the Microbial Succession using a Metagenomic approach in a Barren mined-out site undergoing ecological restoration. For this, he deployed many molecular and computational Biology tool and techniques. He has several publications (articles and book chapters) in International Journals and International Publications. Currently, he is also Chief Editor of Bioinformatics Review and a co-founder/Executive Chief Editor of a Science Journal in Hindi Language "Swadeshi Vigyan" for the native language speaking communities. Both the Journals are a service to the scientific community as they are fully open access.

INTRODUCTION

An invasive species refers to a species that has undergone transportation from its natural habitat to an area where it has been introduced (Vermeij 2005). Such species are referred to as local incidences against a new taxonomic profile; as an invasive organism may invade a territory with success while it may not be able to inhabit another place (Colautti and MacIssac 2004).

An invasion involves three phases:

(1) Dispersal phase that involves transportation of the species from its natural habitat due to changes in climate or a vector or transport.

(2) Lag phase that involves basal levels of the organism in the area. There is a gradual but slow increase in the numbers of the population that are initially below detection. This is followed by the principles of natural selection and adaptation to the new habitat.

(3) Establishment phase involves the successful maintenance of a steady and detectable population. The species can either have a neutral effect on the introduced region or introduce harm in the region (Vermeij 2005).

Literature has pointed out that the success of many invasive species is based on the symbiosis with microbes. The study of the microorganisms that are associated with an organism either externally or internally or the microbiome, reveals that they play key roles in organisms. The aspects of productivity as well as health of organisms ranging from plants, fish, birds, amphibians, insects as well as mammals including humans are influenced by the associated microbes (Kowalski et al., 2015).

The association of such microbes and their hosts covers a broad spectrum of functions as well as highly specific and diverse microbiomes imply that the study of organisms including invasive species requires an understanding of the microbes associated with an organism under study (Gilbert et al., 2012).

Additionally, the study of invasive species and the microbial communities they are associated with as well as the microbes in a native habitat all yield potential ways to target such invasive species for their control (Kowalski et al., 2015).

Due to increase in transport across the globe, invasions are predicted at a larger rate that requires an understanding of the aspects of plant and animal invasions. Studies have revealed that invasive species especially plants alter the microbes in the soil due to shifts in nutrients and elements such as Carbon. Plant invasion have reportedly caused an increase of elements such as nitrogen and carbon which in turn cause an enhancement of microbes in soil (Liao et al., 2008). The interactions between alien and invasive species with that of the native members can aid in the success of such species (Rasool et al., 2016).

An important aspect of ecosystems involves resistance to invasions by alien species (Vitousek et al., 1996). Though the vast repertoire of microbes present in soil display more resistance against invasive species as compared to simple systems, the characterization of equations between invaders and microbes yet require much more study (Jan Dirk van Elsas et al., 2012).

ROLE OF MICROBES AGAINST BACTERIAL INVADERS

The quantity of microbes in soil can be a staggering amount of especially bacteria that play key roles in the functions of soils. The diversity of microbes in soil was found to inhibit an invasive pathogen: *Ralstonia solanacearum* from attacking and causing wilting disease in tomato but this result was not observed in potato. In order to further study the effect of diverse microbes in soil on invasive pathogens, a team in 2012 examined the effect of a derivative of *Escherichia coli* (EHEC)O157:H7 that is associated with haemorrhagic manifestations in soil samples. It was previously studied that in fumigated soils, the diversity of microbes was less that led to persistence of this pathogen. The mechanisms of the resistance to such invasion has been attributed to availability of resources such as organic matter or nutrients in the soil for the pathogen and also communities with lower numbers of microbes had a higher susceptibility to invasion by such pathogens.

Following isolation of soil samples, the bacterial communities were assembled and their effect on the growth of a *E. coli* O157:H7 (strain T) was checked. It was observed that the invasive pathogen (here, strain T) displayed highest survival in sterile soil controls and the lowest survival in natural soil samples. The invader was more in numbers in samples that has 5 strains of microbes as against samples that possessed 20 or 100 microbial strains. Following inoculation, the ability of such invader bacteria Strain T was found to form lesser cfu numbers that showed that the survival of the invader was affected by the diversity of microbes in a sample (using dilution-to-extinction). The survival of the invader was inversely proportional to the diversity of microbial populations used: that included bacteria, fungi, pseudomonads, and Actinobacteria. A greater degree of invasion check was observed for Actinobacteria as well as pseudomonads and bacteria. The maximum value of resistance was observed with Actinobacteria as they are producers of secondary metabolites that influence the success of an invasion.

The mechanism of invasion could be stalled by competition between the invasive bacterial species and the native species for natural resources. The niche theories regarding resources pinpoint the rate at which the invaders and native members compete for resources as well as the amount of resources left for the invasive species following utilization by the native species. Studies for the resource utilization revealed that as the diversity of the native microbial population increased, their ability to use resources increased as compared to the invader (here, strain T) which in turn decreased the competitive ability of the invader.

Thus, in a nutshell these studies conducted showed that the ability of invasive species was compromised in the presence of a native population that had a large diversity of microbes (Jan Dirk van Elsas et al., 2012).

INTERACTIONS BETWEEN PLANTS AND MICROBES

The microbes associated with plants are transmitted to a developing plant from the parent plant or can be acquired through horizontal transmission. The microbes associated with plants may be present on several organs of the plant ranging from endophytes located within a plant or epiphytes present on the surfaces. There are reports of the key roles of the microbes in the invasion of a species: with microbes either playing a positive or negative role in controlling the invasion of the species. As the exact function of the microbes and their symbiotic roles require further clarity in the invading property of a species (van der Putten et al., 2007), the studies of such nature shall cast a light in better management of invasive species.

INVADERS AND MICROBES: TYPES OF INTERACTIONS

The invasion of a non-native plant species (exotic species) has been reported as a large challenge across the world. Over 700,000 hectares of habitats of wildlife are subject to invasion by invasive species annually. There are also reports that invasive species can induce pressure to threaten 400 of 958 threatened or endangered species (Kowalski et al., 2015). It is of significance that to target the control of invasive species involved the expenditure of US $2.2 billion in 2012 by the National Invasive Plants Council (National Invasive Species Council,2014). The following sections shall discuss the roles of several microbes and invasive species:

Native Pathogens and Invaders

Invading species show a preference for areas that have lower environmental factors working against them as well as lower numbers of predators. Darwin had stated that plants and animals record a spurt in population in a new area due to lack of their predators or natural enemies and called it as enemy release hypothesis (ERH). Plants categorized as invaders have been reported to possess pathogens on their roots and foliage when they occupy their original habitat as compared to their new area of invasion.

For invaders in a region, the pathogens present in the natural habitat are non-adapted and hence, are not affected to such an extent as compared to the native plants of the region (Kowalski et al., 2015).

Novel Pathogens and Invaders

According to the biotic resistance hypothesis , for an invading species, a native pathogen of the habitat may serve detrimental and prevent its invasion before it occupies the area to an extent of the native plants endemic to the region .The biotic resistance has been observed in cases when an invading and native species share a close relation; within a region, the pathogens attacked clovers to the same extent when clovers were introduced in a study as the native clovers of the region (Parker and Gilbert, 2004). This is suggestive of pathogens not targeting an invasive species that is of a different taxonomy than native species; however, the numbers of pathogens have shown a rise. A very early study showed that the level of fungi that grew on introduced British trees were an equivalent number to that seen on native trees more than three centuries old (Strong and Levin ,1975).

The probability of a pathogen yielded from successful invasion is very high that can lead to epidemic proportions. Such invasive species may be subject to regulation by pathogens. For instance, *Microstegium vimineum* or Japanese stilt grass; a successful invader was attacked by a microbial fungus: *Bipolaris* to cause a large reduction in its reproduction (Flory et al., 2011).

Mutualist Microbes Native to an Area and Invaders

Many plants are associated with several mutualistic associations with fungi or arbuscular mycorrhizal fungi or Nitrogen fixing bacteria. For an invasive species, if an invader possesses an obligate mutualistic microbial association, the invasion may not be successful. For instance, studies in legumes that possess Nitrogen fixing bacteria have shown that they were not successful in invading an area without their microbes. Another study showed that pines could not be cultivated in Australia until the associated mycorrhizal fungi were also involved (Allen, 1991).

The plants that enter mutualism with microbes and are not transmitted by seeds can successfully invade a habitat as seen by a tree; faya (*Myrica faya)* that has invaded Hawai due to its mutualism with *Frankia*. The association apart from aiding the invasion also plays a key role in Nitrogen metabolism (Walker and Vitousek, 1991).

Novel Mutualism between Microbes and Invasive Species

Invasive species that enter new habitats without the microbes they are in mutualistic associations can also enter into new mutualistic relations with microbes. A microbe: *Balansia cyperi* is endemic to Central and South America entered into a mutualistic association with *Cyperus rotundus*: purple nutsedge a plant endemic to Asia. However, this novel mutualism led to the plant to earn the title of one of the top-ranking weeds: making it a successful invasive species. The transmission is attributed to a shift of *Balansia cyperi* that had infected *Cyperus* species to *Cyperus rotundus* populations that first invaded the gulf coast of the US. Based on the relationship of the native plant and invasive species, such relations may come into force (Kowalski et al., 2015).

SOIL MICROBIAL COMMUNITIES AND INVADING PLANTS

The exact effects of the roles of microbes in the soils in controlling the effects of invasive plants is yet to be elucidated in detail due to limitations in comprehending the speices of microbes involved and their pathways. For instance, an invader, *Aegilops triuncialis* caused a shift in microbial communities into that of a herbaceous one that in turn did not affect any other native species except *Lasthenia californica*.

Mycorrhizal fungi are the initial colonizers of an invasive plant species that in turn cause changes in the levels of other mycorrhizal fungi and native plants in the environment. Several factors are involved with these changes such as the features of the fungi, invaders and natives in the environment along with the conditions. For instance, invasive species, *Carduus pycnocephalus* causes a dip in the numbers of mycorrhizal fungi in the soil which in turn results in a hampered growth of a native plant; *Gnaphalium californicum*. Certain other invasive species utilize the mycorrhizal fungi available in an environment to harness nutrients in order to effectively colonize a new habitat. Additional reports have revealed that certain invasive species can induce alterations in the populations of mycorrhizal fungi in the native environment to suit the needs of the invaders. For instance, *C. maculosa* in Montana, USA and annual grasses in California can induce shifts in the existing population of fungi to suit their requirements. The exact pathways associated with the alterations in such communities yet remain to be classified, there are reports that such invasive species may associate with fungi that facilitate their survival which in turn may give priority to enable the survival of the availed fungal species over others present in the native habitat. The use of substances that inhibit such mycorrhizal fungi of the soils have also been reported (Eviner and Hawkes, 2012).

MICROBIOLOGICAL ANGLES OF WEEDS AND CROPS

Invasive species such as weeds enter into competition with plants and can cause significant damage to ecosystems. It has also been reported that microbes that associate with weeds impart resistance to pathogens as well as growth. The properties of invaders such as weeds are determined by soil microorganisms that can control the competition between the invaders and natives (Massenssini et al., 2014a). The mutualistic association between plants and arbuscular mycorrhizal fungi has the potential to aid in efficient distribution of nutrients in plants to facilitate the existence together while other microbes can influence the density of plants in an area (Massenssini et al., 2014b).

The composition of the microbes associated with a plant can either have a neutral or negative or positive role in the growth of such plants and hence the species of plants in an area. Given the broad spectrum of associations between plants and microbes, the equations between soil microbes and plants can determine the successful invasion of a species (Klironomos, 2002).

The role of mycorrhizal associations can confer the ability of establishment of a species or impart an edge in the competition of a species over another. Reports indicate the formation of arbuscular mycorrhizae between fungi (of which a majority belong to Glomeromycota) and more than 80% of plant species. Given the sheer numbers of the plant numbers involved located spatially across the world, the role of mycorrhizal associations is an important angle (Wang and Qiu, 2006). The arbuscular mycorrhizae are involved in the utilization and allocation of key nutrients such as nitrogen and phosphorus influencing the members of a plant community.

The numbers of arbuscular mycorrhizae were found to be lesser with an invasive species: *Centaurea maculosa* as compared to the diversity associated with the native species (Mummey and Rillig ,2006). Thus, invasive species are thought to alter the levels of such key associations to gain an edge over native species as indicated by this study.

An increase in the ability of a plant to compete with others is conferred by plant growth promoting rhizobacteria or PGPR as these organisms have been shown to increase the ability of plants to resist pathogens and also facilitate uptake of nutrients. As these PGPR also are involved in producing plant hormones, the association of plants with PGPR can impart an edge for competition over other species.

In environments with challenges of phosphorus limitation, an edge over other plants for invasion can be enhanced by associations with Phosphate-solubilizing microorganisms. Another important class of microbes includes: dark septate endophytes found generally with roots of plants and occasionally along with arbuscular mycorrhizae. There are reports of such organisms associating with plants during conditions of stress in the environment.

This large group of microbes that constitute the soil microbiome associated with plants is sensitive to the interactions between invasive species and native plants. There are shifts in the communities of microbes observed when plants are grown solo against when grown together with competitive species (Massenssini et al., 2014a).

INVASIVE WEEDS AND SOIL MICROBES

Weeds are considered to possess properties of invasive species (Reinhart and Callaway, 2006) as they possess features of the extraction of nutrients from the soil coupled with a competitive edge over other species. The interactions of invasive species and microbes may involve another aspect as compared to native species such as crops. Studies were performed on soil that was fumigated (to remove fungi and probable associations with plants including arbuscular mycorrhizal fungi). It was found that while crop plants tested showed an enhanced growth on the soil while the growth of half of invasive weed species tested were hampered in these soils (Santos et al., 2012). This is suggestive of a greater level of associations between soil microbes and such invasive species to form mutualistic associations for their successful growth.

Such invasive species such as weeds may show positive associations with arbuscular mycorrhizal fungi with a key role of the surrounding environment influencing such associations. The mutualistic associations of two species of invaders: *Bidens pilosa* and *Eleusine indica* with arbuscular mycorrhizal fungi was enhanced in the presence of a native plant of a field study: maize. It was hypothesized that the mutualistic association was enhanced to confer an edge to the invasive species (Fialho , 2013).

Studies in literature have pointed out the role of alterations in the communities of microbes associated with invasive species during the colonization or invasion of a new habitat. The accumulation of parasites or pathogens with the invasive species can prove to be detrimental to the invasion while the role of arbuscular mycorrhizal fungi or nutrient fixing bacteria may be beneficial (Reinhart and Callaway, 2006).

In order to have an edge over native species, invasive species may engage in methods of circumventing attack of a pathogen. For instance, the Mediterranean basin was successfully invaded by several species of *Carpobrotus* as the new habitat possessed lesser numbers of pathogens or microbes that entered negative associations with the invaders (Van Grunsven et al., 2009). Thus, the success of the invasive species was traced to its ability to escape from soil pathogens.

Invasive plants can also rely on the use of allelopathic molecules in order to successfully colonize a territory. For instance, a study in 2009 observed that the plant *Solidago canadensis* L was able to invade habitats by the production of such molecules in its root exudates (Zhang et al., 2009). These molecules are implicated in the inhibition of pathogens apart from the controlling the levels of arbuscular mycorrhizal fungi available to the dependent native species, thereby gaining an edge for successful invasion. The invasive members may also cause a shift in the members of arbuscular mycorrhizal fungi in a new terrain such that the growth of species that support the invaders is facilitated. Invasive species are thus capable of altering microbial species in a new environment to enable their growth: to transform potentially negative associations such as pathogens and enhance the growth of species that favour their growth (Massenssini et al., 2014b).

MICROBES AND ANIMAL INVADERS: LIONFISH AND ASSOCIATED MICROBES

The last period of more than two decades has observed a surge in the numbers of a poisonous fish: *Pterois volitans/miles* or Lionfish to the Atlantic Coast and the Caribbean Sea. Figures have been listed as the numbers of invasive fish occupying a fivefold jump in occupying a broader zone as compared to their native locations. These organisms possessed the hallmarks of successful invasive species. In order to examine the microbes associated with this invader, a study in 2014 aimed to examine the communities of microbes.

The techniques of T-RFLP and clone library analyses on the invasive species of lionfish and native fish of the Caribbean Sea: squirrelfish revealed that there were variations in the bacterial communities present on the lionfish and the native fish studied. An interesting point is all the samples were collected at the same time suggestive of exposure to an equivalent poll of bacteria they are exposed to. While *Achromobacter* spp was isolated from only the native squirrelfish; the invasive lionfish showed a larger number (more than fourfold) of *Pseudoalteromonas.* This microbial population has been reported as a producer of metabolites against pathogens such as bacteria and virus thus, conferring a protection to the invasive lionfish against pathogenic attack (Stevens and Olson, 2013) giving it a competitive edge.

As invasive species are known to harbour non-native microbes into the new habitat that confers on them a selective edge such as in this case, microbes on the surface of fish skin that confer protection against pathogens. It was observed that of from 47 lionfish samples, 149 bacterial isolates were cultured of which 54 isolates showed inhibition against a known pathogen: *V. parahaemolyticus.* Analysis of RNA revealed that the microbes that produced these anti-pathogen metabolites were traced to an *Alteromonas* sp. and *Bacillus* spp along with *Micrococcus* sp. and proteobacteria.

There were previous observations that native fish possessed certain fish pathogens on their skin while the invasive lionfish lacked such pathogens on their surface. An invasive species has an important character of retaining symbionts with which it is associated in order to develop resistance against pathogens in the area it invades (Vermeij 2005).

The presence of a microbial community with the lionfish that produces such metabolites against pathogens is thought to confer protection from these pathogens (Julia L. Stevens, 2014).

CASE STUDIES

Case study 1: Alterations in Soil Microbes in Kashmir Soils due to an Invasive Plant Species

A member of the sunflower family: *C. sumatrensis*, initially from South America, has been reported as an invader in several regions such as Kashmir. Apart from the qualities of this plant that include copious amounts of seeds, its resistance to chemical control, the role on microbes has also been investigated. There is a larger diversity of arbuscular mycorrhizal fungi in the roots of *Conyza sumatrensis* when it is in natural habitat as compared to new habitats.

A study in 2016 evaluated to study the effect of such an invader on the profile of various microbes present in the soil of urban ecosystems of the Kashmir. The soil samples that had invaded or non-invaded soil samples were subjected to DNA extraction, PCR amplification and then Denaturing gradient gel electrophoresis. The team reported the presence of 57 different species of bacteria with major representatives as Sphingobacteria and Alphaproteobacteria. The former bacterium was present in a ratio of 16 % in the un-invaded plots while 34 % in invaded plots. The plots invaded by the plants showed higher levels of Cyanobacteria while Acidobacteria were found in un-invaded plots. 13 % of the betaproteobacteria were reported in plots in the invaded fields while a lower level (8%) in fields without invasive species.

Thirty-five different species of fungi were isolated of which more numbers of Ascomycetes were recorded in the plots that had the invader. Species of *Peziza subcitrina* (Pezizaceae), *Paraphoma chrysanthemicola* (Phaeosphaeriaceae), and Penicillium sp along with members of *Phaeosphaeriaceae*, *Trichocomaceae*, and *Ascobolaceae* were reported only from the samples that had been invaded. For instance, *Pleosporaceae* contributed 20 % of the plots invaded as compared only 13 % in the un-invaded plots.

The results together showed that the plant invasive species: *Conyza sumatrensis,* introduces a shift in microbial populations of soil in Kashmir. In a nutshell, the invader decreases the diversity of the soil while on the other hand; the organism causes a specialization of organisms that were exclusively. The nature of the native plants along with the features of the invaders all influence the composition of the microbes associated. *Conyza* is associated with the production of several allelopathic moleculues that include phenol based compounds. These phenolic compounds cause changes in nutrient fluxes and alterations in membrane permeability that can lead to a decrease in numbers of associated microbes especially bacteria, or even cause their wipe out. Fungi have been shown to use phenolic compounds as a carbon source hence, in this study a greater enrichment of fungi were reported in samples that were invaded by the plant (Reviewed by Rasool et al., 2016).

Case study 2: Microbes and Reed Invaders

In order to target invasive species, researchers are targeting the microbes associated with an invader. The following study shall highlight such an approach:

Common reed: A native species of wetlands of North America is *Phragmites australis* spp. *americanus*; that is being actively replaced by an invader from Europe: *Phragmites australis* spp. *Australis.* The invasion of the plant results in a reduction of biodiversity of the area. An expenditure of $4.6 million as incurred in the use of chemicals or mechanical approaches to combat the reed between 2005 and 2009. The current management of this invader involves application of herbicides as well as burning and chopping. However, the approach has been reported as involving a lot of time and expenditure apart from the side effects of usage of herbicides.

The microbial communities involved with *Phragmites* include a range of genera such as with endophytic fungi such as *Stagonospora* contributing to its growth. Other microbes associated include water molds or oomycete, bacterial and archaeal epiphytes as well as fungi. A study evaluated the ability of endophytic bacteria associated with the reed in the detoxification of polluted waters; suggesting that these bacteria confer on the plant to invade even contaminated water (Chen et al., 2012).

A novel strategy has been designed for addressing the invasive organism through modification of the microbial communities associated with *Phragmites*. The Collaborative for Microbial Symbiosis and *Phragmites* Management established by the Great Lakes Commission and U.S. Geological Survey to apply an amalgamated technology against the organism. The overall strategy of the management system involves the following steps to alter the microbes associated with *Phragmites:*

- *Data on the microbes*: This involves a comparison of the microbes associated with native and invasive species, apart from characterization of the endophytes in the invader.
- *Microbial benefits*: This stage entails the effect of various inoculants on the growth of *Phragmites* apart from pathogens or microbes that increase competition in the habitat.
- *Manipulating microbial associations*: In this stage, inhibitors (that have very little effect on the environment) to microbes associated with the plant are evaluated and their effects when grown along with the native species (Reviewed by Kowalski et al., 2015).

CONCLUSION

Invasions by several organisms that include bacteria, plants and animals have been well documented. A comprehensive list of invasive species has been listed at the US Department of Agriculture: https://www.invasivespeciesinfo.gov/index.shtml. Certain invasive plants include Air Potato (*Dioscorea bulbifera*), Cogongrass (*Imperata cylindrica*), Russian Knapweed (*Rhaponticum repens*) etc. while invasive species of animals include: Burmese Python (*Python molurus bivittatus*), Glassy-Winged Sharpshooter (*Homalodisca vitripennis*),

Asian Long-Horned Beetle (*Anoplophora glabripennis*) etc. Additionally, microbes that are invasive include viruses such as Zika Virus Disease, West Nile Virus (*Flavivirus*) or pathogens such as Citrus Greening (*Liberibacter asiaticus*), White Pine Blister Rust (*Cronartium ribicola*) etc. in the US.

This chapter examined the role of microbes in the establishment or failure of an invasive species using examples from bacterial, plant as well as animal kingdoms. Microbes have been reported to play a significant role in establishing an invasive species successfully in a new habitat either by conferring resistance to pathogens or enable its survival by aiding uptake of nutrients. These aspects of the study of microbes and invasive species can aid in the designing of targets against such invasions.

REFERENCES

1. Allen MF. The Ecology of Mycorrhizae. Cambridge: Cambridge University Press. 1991.
2. Chen W-M, Tang Y-Q, Mori K, Wu X-L. (2012). Distribution of culturable endophytic bacteria in aquatic plants and their potential for bioremediation in polluted waters. Aquat Biol 15: 99– 110.
3. Colautti RI, HJ MacIsaac. (2004). A neutral terminology to define 'invasive' species. Divers Distribution 10:135-141.
4. Elsas JD, Chiurazzi M, Mallon CA, Elhottova D, Krištu° fek V, Salles JF. (2012). Microbial diversity determines the invasion of soil by a bacterial pathogen. Proc Natl Acad Sci 109 (4): 1159–1164.
5. Fialho CMT. Interação entre micro-organismos do solo, plantas daninhas e as culturas do milho e da soja. 2013. 75 f. Tese (Doutorado em Fitotecnia) – Universidade Federal de Viçosa, Viçosa, MG, 2014.
6. Flory SL, Kleczewski N, Clay K. (2011). Ecological consequences of pathogen accumulation on an invasive grass. Ecosphere 2: 120. doi:10.1890/ ES11-00191.1.
7. Gilbert SF, Sapp J, Tauber AI. (2012). A symbiotic view of life: we have never been individuals. Q Rev Biol 87: 325–341. doi:10.1086/668166.
8. Klironomos JN. (2002). Feedback with soil biota contributes to plant

rarity and invasiveness in communities. Nature 417(6884): 67-70.

9. Kowalski KP, Bacon C, Bickford W, Braun H, Clay K, Leduc-Lapierre M, et al. (2015). Advancing the science of microbial symbiosis to support invasive species management: a case study on Phragmites in the Great Lakes. Front Microbiol 6 (95): 1-14.
10. Liao C, Peng R, Luo YQ, Zhou XH, Wu XW, Fang C, et al. (2008). Altered ecosystem carbon and nitrogen cycles by plant invasion: a meta-analysis. New Phytol 177:706–714.
11. Massenssini AM, Bonduki VHA, Melo CAD, Tótola MR, Ferreira FA, Costa MD. (2014b). Soil Microorganisms And Their Role In The Interactions Between Weeds And Crops. Planta Daninha, Viçosa-MG 32, (4): 873-884, 2014b.
12. Massenssini AM. et al. (2014a). Arbuscular mycorrhizal associations and occurrence of dark septate endophytes in the roots of Brazilian weed plants. Mycorrhiza 24, (2): 153-159.
13. Mummey D, Rilling M. (2006). The invasive plant species Centaurea maculosa alters arbuscular mycorrhizal fungal communities in the field. Plant Soil 288 91): 81-90.
14. National Invasive Species Council. Invasive Species Interagency Crosscut Budget. 2014; Available online at: http://www.invasivespecies.gov/global/orgcollab budget/ orgcollabbudgetdocuments/NISC2013CrosscutBudgetSummary.pdf
15. Parker IM, Gilbert GS. (2004). The evolutionary ecology of novel plant- pathogen interactions. Annu Rev Ecol Evol S 35: 675–700. doi:10.1146/ annurev. ecolsys.34.011802.132339.
16. Rasool N, Reshi ZA, Khasa DP, Roshan M, Shah MA. (2016). Invasion by Conyza sumatrensis alters soil microbial community structure in urban ecosystems. Ecol Process 5:10. DOI 10.1186/ s13717-016-0054-4.
17. Reinhart KO, Callaway RM. (2006). Soil biota and invasive plants. New Phytol 170 (3): 445-457.
18. Santos EA, Ferreira, Costa LR, Santos MD, Silva JB, Soares da MC, Ignacio A. (2012). The effects of soil fumigation on the growth and mineral nutrition of weeds and crops. Acta Sci Agron 34 (2): 207-212.

19. Stevens JL, JB Olson. (2013). Invasive lionfish harbor a different external bacterial community than native Bahamian fishes. Coral Reefs 32:1113-1121.
20. Stevens JL. Assessment Of The Bacterial Communities Associated With Invasive Indo-pacific Lionfish, A Study Of Biogeography And Host-microbe Interactions. Dissertation. 2014. Department of Biological Sciences in the Graduate School of The University of Alabama.
21. Strong DR, Levin DA. (1975). Species richness of the parasitic fungi of British trees. Proc Natl Acad Sci USA 72: 2116–2119. doi:10.1073/ pnas.72.6.2116.
22. Valerie T. Eviner and Christine V. Hawkes. The Effects of Plant–Soil Feedbacks on Invasive Plants: Mechanisms and Potential Management Options. CAB International 2012. Invasive Plant Ecology and Management: Linking Processes to Practice (eds T.A. Monaco and R.L. Sheley.
23. Van derPutten WH, Klironomos JN, Wardle DA. (2007). Microbial ecology of biological invasions. ISME J 1: 28–37.doi:10.1038/ ismej.2007.9
24. Van Grunsven RHA, Bos F, Ripley BS, Suehs CM, Veenedaal EM. (2009). Release from soil pathogens plays an important role in the success of invasive Carpobrotus in the Mediterranean. South African J Bot 75 (1): 172-175.
25. Vermeij GJ. Invasion as expectation: a historical fact of life. In Species Invasions: Insights into Ecology, Evolution and Biogeography. 2005; (Sax, DF, et al., eds), pp.315-340, Sinauer Associates.
26. Vitousek PM, Dantonio CM, Loope LL, Westbrooks R. (1996). Biological invasions as global environmental change. Am Sci 84:468– 478.
27. Walker LR, Vitousek PM. (1991). An invader alters germination and growth of native dominant tree in Hawaii. Ecology 72, 1449–1455.doi:10.2307/1941117.
28. Wang, Qiu. (2006). Phylogenetic distribution and evolution of mycorrhizas in land plants. Mycorrhiza 16 (5): 299-363.

29. Zhang SY, J, Chen X. (2009). The invasive plant Solidago canadensis L. suppresses local soil pathogens through allelopathy. Appl Soil Ecol 41(2): 215-222.

7

CHAPTER

TOWARD A CENSUS OF BACTERIA IN SOIL

Patrick D. Schloss, Jo Handelsman

Department of Plant Pathology, University of Wisconsin–Madison, Madison, Wisconsin, United States of America

ABSTRACT

For more than a century, microbiologists have sought to determine the species richness of bacteria in soil, but the extreme complexity and unknown structure of soil microbial communities have obscured the answer. We developed a statistical model that makes the problem of estimating richness statistically accessible by evaluating the characteristics of samples drawn from simulated communities with parametric community distributions. We identified simulated communities with rank-abundance distributions that followed a truncated lognormal distribution whose samples resembled the structure of 16S rRNA gene sequence collections made using Alaskan and Minnesotan soils. The simulated communities constructed based on the distribution of 16S rRNA gene sequences sampled from the Alaskan and Minnesotan soils had a richness of 5,000 and 2,000 operational taxonomic units (OTUs), respectively, where an OTU represents a collection of sequences not more than 3% distant from each other. To sample each of these OTUs in the Alaskan 16S rRNA gene library at least twice, 480,000 sequences would be required; however, to estimate the richness of the simulated communities using nonparametric richness estimators would require only 18,000 sequences. Quantifying the richness of complex environments such as soil is an important step in building an ecological framework. We have shown that generating sufficient sequence data to do so requires less sequencing effort than completely sequencing a bacterial genome.

SYNOPSIS

Soil is more than dirt. It is the source and sink of nutrients, wastes, pharmaceuticals, and energy required to make Earth supportive of life—it is Earth's most vital organ. Although we know a considerable amount about the physical structure and chemistry of soil, there is a glaring paucity of knowledge regarding the microbial component responsible for its many functions. Over the past 100 years, microbiologists have attempted to characterize the biodiversity of microbial life in soil, and many had reached the unsatisfying conclusion that bacteria may be too diverse to count. Schloss and Handelsman have developed statistical models that they apply to molecular data to predict that the richness of bacteria in 0.5-g soil samples from Alaska and Minnesota were 5,000 and 2,000 species, respectively. At the current level of sampling, approximately 20% of the bacteria appear to be endemic to both soils. The enumeration and description of these organisms points to the need and relative ease of characterizing bacterial communities to identify the organisms responsible for sustaining all of life.

INTRODUCTION

Enumerating the human population of a country or region through a census is an ancient problem that is complicated by the challenges inherent in accurately representing a large and often inaccessible population.

The same issues manifest in censuses of microbial communities, but are intensified by greater complexity and methodological challenges. Although a complete census of a country is theoretically possible, it is currently impractical to survey all 10^9 bacterial cells in a gram of soil [1], making a sample-based census the best option for estimating richness—the number of bacterial taxa in soil. To do so accurately requires a reliable means to access the bacteria, a reasonable definition of "species," and a robust description of the frequency distribution of the species. Just as a country›s census describes a fundamental property of that country, an environment›s richness is the most fundamental descriptor of community structure, and patterns of richness can be correlated with an environment›s geography, productivity, extremeness, climate change, and degree of isolation [2]. Our inability to estimate richness impedes investigation of the effects of soil chemistry, pollution, and land use on the soil microbial community.

The method used to access the microbial biodiversity assuredly shapes the outcome of a census. Culture-based methods suggest that a gram of soil contains fewer than 100 species [3], but these are undoubtedly underestimates because multiple lines of evidence indicate that fewer than 1% of the species in soil are presently culturable [4]. Culture-independent methods include DNA reassociation and 16S rRNA gene sequencing, which have provided conflicting results due to the problems inherent in defining a species and in estimating the frequency distribution of species in soil. Depending on how the data are analyzed, DNA reassociation experiments produce richness estimates ranging from 4,000 to 10,000,000 genome equivalents per 10 or 30 g of soil [5–11]. The variability in these estimates stems from application of different assumptions to reassociation curves, and their interpretation is complicated by the lack of controls that account for intergenomic variation. Finally, DNA reassociation kinetics cannot be used to compare the membership of different communities.

An alternative method relies on analysis of 16S rRNA gene sequences amplified from soil by PCR [12]. The power of this method lies in its use of the universal tool of bacterial phylogeny and our ability to define operational taxonomic units (OTUs) based on the relatedness of sequences. Estimates of richness have been obtained through parametric or nonparametric empirical models of species frequency distribution to produce richness estimates between 590 and 100,000 species per gram of soil [13–15]. Parametric models have assumed that the incidence of different species follows a lognormal [13], Pareto [16], or uniform distribution [14]. Although the lognormal model has been useful as a "null model" [17], data are insufficient from any soil community to support reliance on a lognormal or Pareto frequency distribution, and we are unaware of any dataset that supports a uniform frequency distribution [18]. Analyses based on nonparametric models, which do not assume a defined frequency distribution but are based on the frequency of abundant community members [19–21], estimate a minimum richness of 590 species based on 16S rRNA gene sequences in a Scottish soil [15,22]. Although the extent of the universality of phylum-specific PCR primers and potential toxicity of some fragments is not well understood, these effects would reduce the perceived richness. Finally, use of 16S rRNA gene sequences permits direct comparisons of the membership of different communities.

Previously, Dunbar et al. [17] modeled the frequency distribution of 16S rRNA genes in four Arizonan soil communities by fitting a lognormal frequency distribution. Using 200 16S rRNA gene fragments from each soil community, this analysis estimated that 10 g of soil contained between 3,000 and 8,000 16S rRNA gene restriction fragment length polymorphism (RFLP) profiles. Previous analysis using the same four libraries found that the similarity of 16S rRNA gene sequences with the same RFLP profile ranged between 52.2% and 99.9% [3], which makes interpretation of the analysis difficult. We were interested in developing this approach further by analyzing large 16S rRNA gene sequence collections that had not been initially screened by RFLP profiling. Our approach was to find a simulated community whose samples resembled our sampling of 1,033 16S rRNA genes from a clone library constructed from a single 0.5-g sample of Alaskan soil. For the purposes of comparison, we also analyzed two large 16S rRNA gene sequence collections that were recently published as part of a soil metagenomic sequencing project, but were not characterized beyond their phylogenetic affiliation [23].

RESULTS

Estimating the Bacterial Richness in the Alaskan Soil Library

The aim of this work was to estimate the taxonomic richness in an Alaskan soil sample through a library of 16S rRNA gene sequences derived from the sample. We assigned more than 92% of the 1,033 Alaskan 16S rRNA gene sequences to seven phyla, including the Proteobacteria (48.6%), Acidobacteria (15.3%), Bacteroidetes (9.3%), Actinobacteria (5.8%), Gemmimonas (5.7%), Planctomycetes (4.0%), and Verrucomicrobia (4.0%); the remaining sequences clustered within 12 phylum-level delineations, four of which had no cultured members (Figure 1). Each phylum was sampled at least twice, except for candidate phylum BD Group and the phylum Chlamydiae, which were each observed once.

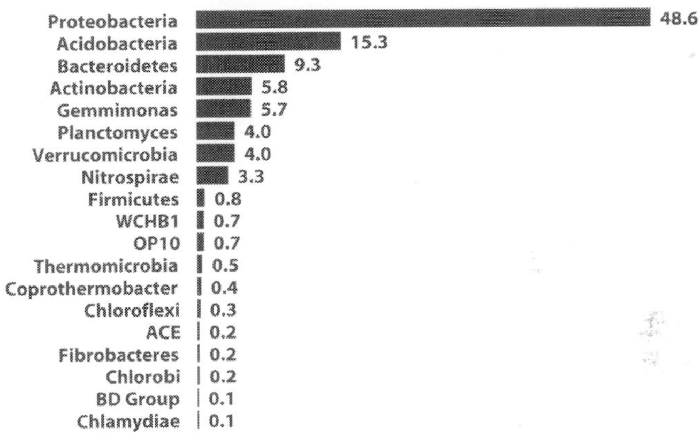

Figure 1: Phylum-Level Delineation of the 16S rRNA Gene Fragments in Alaskan Soil.

Gene fragments (n = 1,033) were isolated and sequenced from an Alaskan soil. Candidate phyla WCHB1, OP10, ACE, and BD Group have no sequenced representatives.

https://doi.org/10.1371/journal.pcbi.0020092.g001

We used furthest neighbor clustering to assign sequences to OTUs based on the pairwise genetic distance between sequences. Although controversial [24], Jukes-Cantor-corrected distances less than 0.03 are considered to correspond to a strain-level delineation, 0.03 to species, 0.05 to genus, 0.15 to class, and 0.30–0.40 to phylum [25–28]. Considering potential intragenomic differences between copies of 16S rRNA genes and errors due to sequencing and alignment [29,30], the 0.03 cutoff is also a pragmatic choice since it probably represents the most stringent OTU definition that is practically obtainable using 16S rRNA genes. Since the intragenomic distance between 16S rRNA gene sequences is typically less than 0.03, at this distance, replicate 16S rRNA gene sequences from the same genome would form a single OTU.

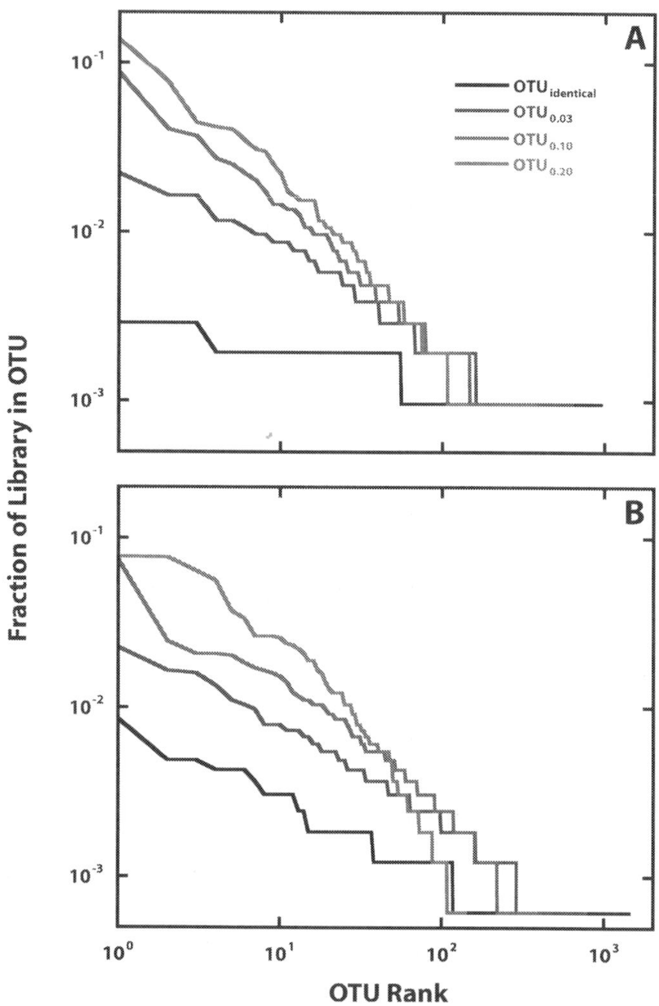

Figure 2: Rank Abundance Plot of the Alaskan and Minnesotan 16S rRNA Gene Libraries.

Alaskan (n = 1,033) (A) and Minnesotan (n = 1,633) (B) 16S rRNA gene libraries are plotted and describe the distribution of the 16S rRNA genes among OTUs defined as a group of sequences that are either identical or no more than 3%, 10%, or 20% different.

https://doi.org/10.1371/journal.pcbi.0020092.g002

To simplify the reporting of our results, OTUs will be designated $OTU_{x.xx}$, where the subscript represents the maximum distance between any two sequences within that OTU (Figure 2). In the Alaskan 16S rRNA gene sequence collection containing 1,033 sequences, we observed 633 $OTUs_{0.03}$. We observed 472 $OTUs_{0.03}$ once and 94 $OTUs_{0.03}$ twice (Figure 2A). The three most abundant $OTUs_{0.03}$ affiliated with members of the phylum Gemmimonas (n = 23 sequences in the $OTU_{0.03}$ from 19 distinct primary sequences), Duganella sp. (n = 17 sequences in the $OTU_{0.03}$ from 13 distinct sequences), and Rhodoferax sp. (n = 17 sequences in the $OTU_{0.03}$ from 15 distinct sequences). These three $OTUs_{0.03}$ were not observed in the Minnesotan sequence collection.

Since the most abundant $OTU_{0.03}$ in the Alaskan 16S rRNA gene library was observed only 23 times, we were unable to obtain meaningful fits of parametric frequency distribution models to the $OTU_{0.03}$ frequency distribution [31]. Attempts to identify parameters that would define simulated communities following either a Pareto or uniform frequency distribution resembling the observed distribution were unsuccessful. The predicted abundance of the most abundant OTUs in the Pareto-distributed communities was too high, and the abundance of the rarest OTUs was too low. We successfully simulated the $OTU_{0.03}$ frequency distribution observed in the Alaskan 16S rRNA gene sequence collection by altering the richness and evenness of random frequency distributions using a truncated lognormal frequency distribution (Table 1). The relative abundance of each OTU in the simulated communities that followed the truncated lognormal model was

Table 1: Example of Simulation Results for Lognormal and Uniformly Distributed Communities with a Richness of 5,000

N_T/N_{max}	Mean (μ)	Standard Deviation (σ)	1/D	Sampling Effort Required for Census			
				Complete	95% of Richness	Chao1 Estimator	ACE Estimator
10	6.000	4.901	64	560,000	85,000	42,000	17,000
	8.000	6.146	66	475,000	74,000	38,000	13,000
	10.00	7.189	67	450,000	70,000	35,000	12,000
35	6.000	3.020	288	480,000	71,000	39,000	18,000
	8.000	3.813	373	305,000	48,000	26,000	9,000
	10.00	4.489	419	203,000	43,000	20,000	8,000
100	6.000	2.481	541	385,000	70,000	40,000	20,000
	8.000	3.153	891	260,000	41,000	21,000	7,000
	10.00	3.719	1,124	200,000	33,000	16,000	5,000
1,000	8.000	2.119	2,680	185,000	29,000	16,000	8,000
	10.00	2.759	3,175	140,000	23,000	11,000	6,000
	12.00	3.232	3,484	120,000	21,000	8,000	5,000
5,000	Uniform		5,000	75,000	15,000	150	110

The sampling effort represents the size of sample necessary to observe every taxon twice, to observe 95% of the taxa, or for the CI of the Chao1 and ACE richness estimators to include the true richness. For each distribution, 1,000 random communities were drawn. Simulation results were positively correlated with richness. N_T represents the total number of individuals in a community and N_{max} represents the abundance of the most abundant member in the community, and their ratio represents the reciprocal of the probability observed at the distribution's mode. The reciprocal of the Simpson's Index (1/D) represents the number of uniformly abundant OTUs needed to observe the same level of diversity found in the community.
DOI: 10.1371/journal.pcbi.0020092.t001

$$N_i = \frac{\frac{1}{S_i\sigma\sqrt{2\pi}}\exp[-\frac{1}{2}\left(\frac{\ln S_i-\mu}{\sigma}\right)^2]}{\int_1^{S_T}\frac{1}{S\sigma\sqrt{2\pi}}\exp[-\frac{1}{2}\left(\frac{\ln S-\mu}{\sigma}\right)^2]dS} \quad (1)$$

where S_i is the ith OTU and N_i is the relative abundance of individuals in that OTU. The maximum possible value of i is the total number of OTUs in the community, S_T. N_1 is the abundance of the most abundant OTU (N_{max}), and N_T is the sum of all N_i values.

Next, we heuristically identified the normal mean (μ = 6.000), standard deviation (σ = 3.020), and $OTU_{0.03}$ richness (S_T = 5,000) for a truncated lognormal distribution in which the distribution of its samples resembled the distribution observed in the Alaskan sequence collection (Figure 2A and Table 1). Further confirmation for the plausibility of the simulated community was obtained by comparing the percentage of the total community represented by the most abundant OTU ($100 \times N_{max}/N_T$) in the simulated community (2.9%) to the value observed from the sequence collection (2.2%). These values are comparable to the range 2.9%–8.3% observed by Dunbar et al. [17], but are considerably higher than the range 0.1%–1% suggested by Curtis et al. [13]. We found that the reciprocal of the Simpson's index (1/D) for the simulated community was 288, which was similar to the value observed for the sequence collection of 223. The values for 1/D are considerably higher than the range 52–107 observed by Dunbar et al. [17].

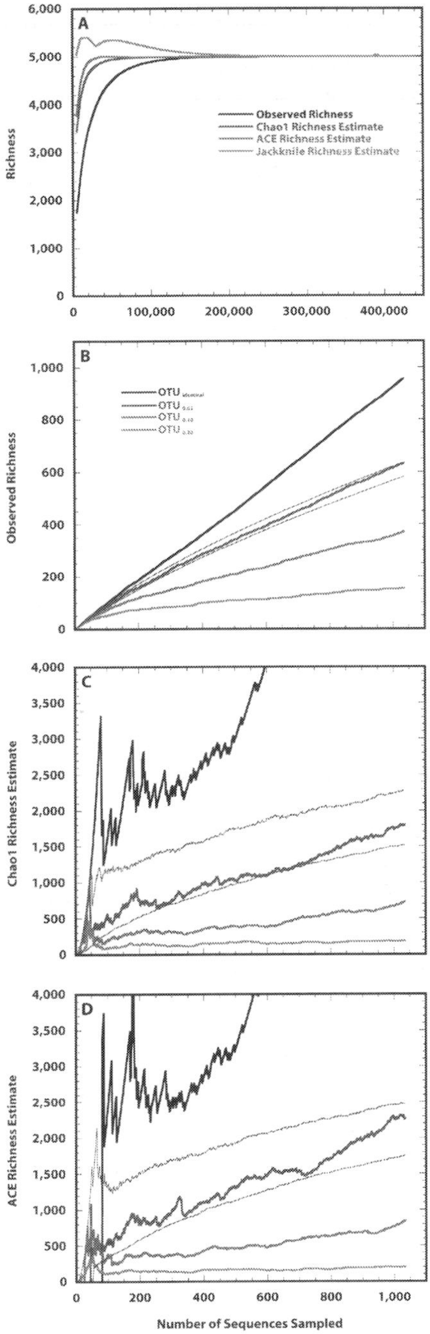

Figure 3: Estimating the Richness of Taxa in the Simulated Alaskan Soil Community.

(A) Average number of taxa observed and estimated richness for the simulated Alaskan soil community ($S_T = 5{,}000$, $\mu = 6.000$, and $\sigma = 3.020$) over the course of randomly sampling 480,000 individuals.

(B–D) The distribution of 16S rRNA sequences obtained from Alaskan soil falls within the 95% CI that would have been obtained for the distribution derived from sampling 1,033 individuals from this simulated community as measured using the observed (B) and estimated—Chao1 (C) and ACE (D)—richness. The thin blue lines in (B), (C), and (D) represent the 95% CIs for each metric using the simulated Alaskan soil community.

https://doi.org/10.1371/journal.pcbi.0020092.g003

To sample every OTU in the Alaskan simulated community twice with 95% confidence would require sequencing 480,000 16S rRNA gene fragments, and to observe 95% of the richness, 71,000 16S rRNA genes would be required (Figure 3 and Table 1). To obtain an estimate of the true richness using either the ACE or Chao1 nonparametric richness estimator would require sampling 18,000 or 39,000 16S rRNA genes, respectively, which represented sampling 65% and 85% of the true richness (Figure 3 and Table 1).

Since we were unable to obtain a robust estimate of species richness with our 16S rRNA gene sequence collection without assuming some distribution a priori, we relaxed the OTU definition to obtain a robust nonparametric richness estimate. The $OTU_{0.20}$ richness estimate collector's curves began to stabilize late in sampling (Figure 3). Although additional sampling would improve the precision of the $OTU_{0.20}$ richness estimate, the Chao1 (188.20, 95% confidence interval [CI] 174–212), ACE (200, CI 181–234), and Jackknife (203, CI 184–222) estimates were similar.

Comparison of Alaskan and Minnesotan Soils Microbial Communities

Recently, the microbial community of a Minnesota farm soil was characterized by metagenomic (direct cloning and analysis of DNA from a soil sample) and 16S rRNA gene sequencing analyses [23]. The authors constructed two separate 16S rRNA gene libraries by using a cell fractionation-based DNA isolation procedure, and sequenced 1,633 overlapping gene fragments from the two libraries [23,32].

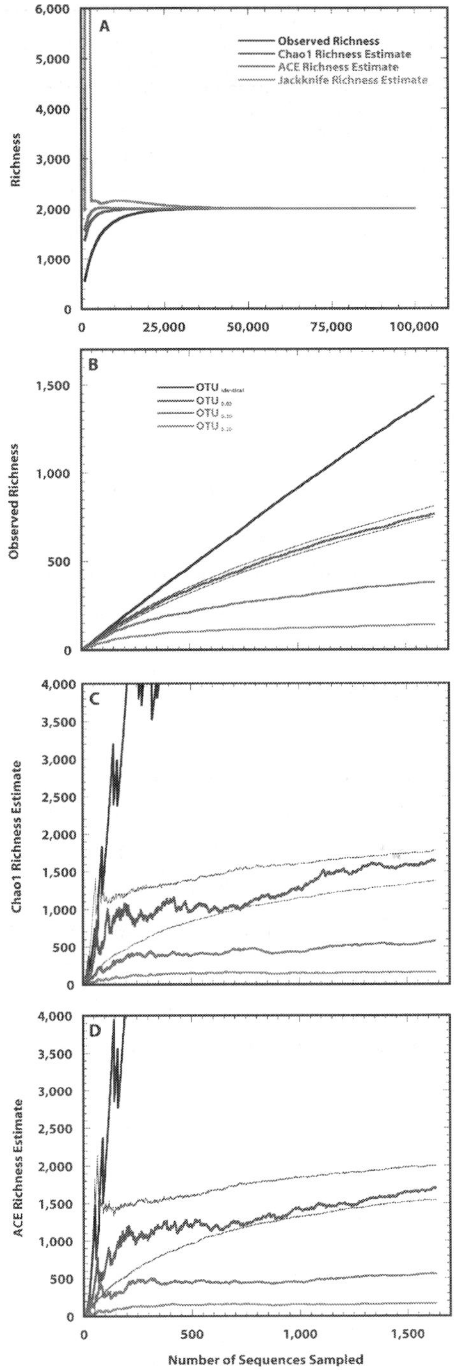

Figure 4: Estimating the Richness of Taxa in the Simulated Minnesotan Soil Community.

(A) Average number of taxa observed and the estimated richness for the simulated Minnesotan soil community (S_T = 2,000, μ = 8.000, and σ = 3.813) over the course of randomly sampling 100,000 individuals.

(B–D) The distribution of 16S rRNA gene sequences obtained from the Minnesotan soil falls within the 95% CI that would have been obtained for the distribution derived from sampling 1,633 individuals from this simulated community as measured using the observed (B) and estimated—Chao1 (C) and ACE (D)—richness. The thin blue lines in (B), (C), and (D) represent the 95% CI for each metric using the simulated Minnesotan soil community.

https://doi.org/10.1371/journal.pcbi.0020092.g004

We reanalyzed their pooled sequence data to determine the richness of the Minnesota farm soil and to determine the degree of OTU membership that was conserved between the Minnesotan and Alaskan soil communities.

Collector's curves for the number of $OTU_{0.03}$ observed and estimated in the Minnesota soil library were flatter than the Alaskan collector's curves (Figure 4). In the Minnesotan collection, the observed $OTU_{0.03}$ richness was 767, and we observed 477 $OTUs_{0.03}$ once and 128 $OTUs_{0.03}$ twice (Figure 2B). The nonparametric richness estimates were 1,647 (Chao1), 1,704 (ACE), and 2,248 (Jackknife); however, each estimate continued to increase with sampling. The three $OTUs_{0.03}$ most frequently observed in the Minnesotan sequence collection contained 37, 27, and 26 sequences, and each clustered within the phylum Chloroflexi; no representatives of these $OTUs_{0.03}$ were observed in the Alaskan sequence collection.

We identified one simulated community with a truncated lognormal distribution that had a richness of 2,000 (μ = 8.000, σ = 3.813) whose samples resembled the distribution observed in the Minnesotan sequence collection. The percentage of the clones represented by the most abundant $OTU_{0.03}$ was 2.3%, and it was 3.9% in the simulated community. The simulated community had a higher 1/D value than that observed from the sequence data (311 versus 237). The Minnesotan simulated community had a lower richness and more uniform evenness than we observed for the Alaskan simulated community. If the Minnesotan simulated community is a true reflection of the $OTU_{0.03}$ distribution, then we are 95% confident that sequencing of 90,000 16S rRNA genes would result in observing every $OTU_{0.03}$ at least twice. Sequencing 16,000 16S rRNA gene fragments would allow us to observe 95% of the true richness. To obtain a nonparametric estimate of richness through the ACE and Chao1 estimators, 2,000 and 5,500 16S rRNA gene sequences, respectively, would need to be sequenced for the estimate›s CI to include 2,000; the CI for the Jackknife estimator already includes 2,000. The sequencing of 2,000 and 5,500 16S rRNA genes would result in observing 43.9% and 72.5% of the true richness, respectively.

It is difficult to determine whether the difference in estimated richness between the Alaskan and Minnesotan simulated communities was due to ecological differences or differences in DNA extraction methods [33] or both. The collector›s curve for the estimated fraction of the Minnesotan library shared with the Alaskan library, 0.18 (standard error [SE] = 0.07), indicated that this value is close to the true value and that the fraction of the Alaskan library shared with the Minnesotan library, 0.17 (SE = 0.06), continued to increase with additional sampling (Figure 5A). Our observation that 18% of the sequences in the Minnesotan library belonged to $OTUs_{0.03}$ shared with the Alaskan collection indicates either that a large fraction of these $OTUs_{0.03}$ are endemic to different soils or that the different DNA extraction procedures preferentially lysed a subset of the $OTU_{0.03}$ membership, or both.

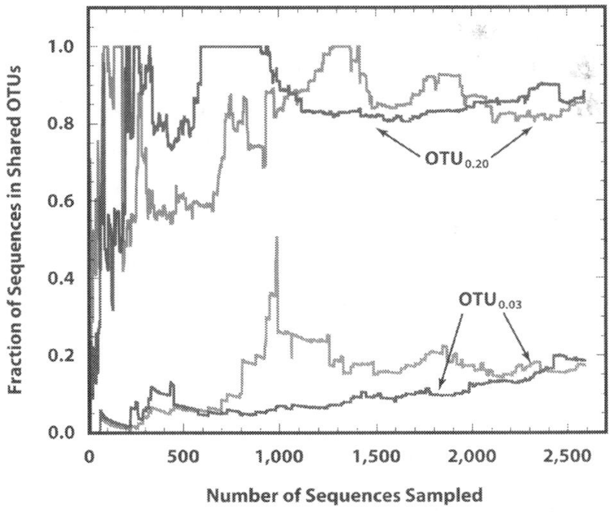

Figure 5: Similarity of Alaskan and Minnesotan Soil Microbial Communities.

Collector›s curves describing the effect of sampling on the estimated fraction of sequences from the Minnesotan (red lines) and Alaskan (blue lines) libraries belonging to shared $OTUs_{0.03}$ and $OTUs_{0.20}$.

https://doi.org/10.1371/journal.pcbi.0020092.g005

Similar to our analysis of the Alaskan 16S rRNA library, when we relaxed the OTU definition to analyze the $OTU_{0.20}$ richness of the Minnesotan 16S rRNA membership, we observed richness estimates that were not sensitive to further sampling. The terminal Chao1 (165, CI 151–197), ACE (169, CI 156–196), and Jackknife (174, CI 158–190) estimates were similar; this is approximately 85% of the $OTU_{0.20}$ richness observed in the Alaskan 16S rRNA library. The fraction of sequences from the Minnesotan library that belonged to $OTUs_{0.20}$ shared between the two libraries was 0.86 (SE = 0.10) and the fraction of sequences from the Alaskan sequences that belonged to $OTUs_{0.20}$ shared between the two libraries was 0.88 (SE = 0.07) (Figure 5B). At this point in sampling, it was not possible to conclude with statistical confidence that the $OTU_{0.20}$ memberships were significantly different, since both CIs included 1.00; however, we expect further sampling to make the estimates more precise.

DISCUSSION

In the 20th century, the view of soil microbial ecology shifted from being described by Selman Waksman as a "clear picture" [34] to E. O. Wilson's pronouncement that its diversity is "beyond practical calculation" [35]. We have shown that neither view is wholly correct, but that a confident estimate of bacterial richness is attainable using a set of parameters that have a reasonable biological basis. We have shown that it is possible to obtain an $OTU_{0.03}$ richness estimate for soil for considerably less effort than is required to shotgun sequence a bacterial genome (assuming ~100,000 sequence reads per genome and one to five reads for each of 17,000 16S rRNA gene fragments). Determining the richness of specific phylogenetic groups using lineage-specific PCR primers would further reduce the required effort.

Our analysis can also be applied to guide the design of functional and sequence-based metagenomics projects [36]. Tringe et al. [23] estimated that more than 2×10^9 bp of sequence from 3×10^6 sequence reads would be necessary to obtain 8-fold sequence coverage of the most abundant species in their soil sample assuming a genome size of 6 Mbp. To sequence 8-fold coverage of the most abundant $OTU_{0.03}$ from the simulated Alaskan soil community, approximately 450 genome equivalents, or 3×10^9 bp, would need to be sequenced from the Alaskan soil. To sequence 8-fold coverage of the ten most abundant $OTUs_{0.03}$ from the simulated Alaskan soil community, approximately 1,600 genome equivalents, or 10^{10} bp, would need to be sequenced. Although this amount of DNA may be beyond our current sequencing capacity, the 10^{10} bp is approximately the content of a 275,000-clone fosmid library. Such a library could be easily constructed and would be useful for functional metagenomic approaches. Although not currently feasible, sequencing 8-fold coverage of every $OTU_{0.03}$ in the Alaskan soil metagenome would require sequencing 950,000 genome equivalents or 6×10^{12} bp of DNA. Although PCR bias may affect the true community distribution, these values are a helpful guide when designing metagenomics-based experiments. For some groups of organisms, the 3% cutoff between 16S rRNA gene sequences has been found to correlate with 70% similarity between genome sequences; therefore, it is unclear how many contigs would assemble for the predicted level of sequencing effort given the substantial intragenomic variation that may exist between members of the same $OTU_{0.03}$.

Estimating richness does not provide the identity of each bacterial type; in the Alaskan soil we studied, identifying every one of the 5,000 different types of bacteria would require sampling more than 480,000 sequences. Furthermore, our analysis assumes an operational species definition of a group of 16S rRNA sequences that are no more than 3% different from one another. Among the members of a single OTU, there is undoubtedly considerable phenotypic and genomic diversity that is not reflected by 16S rRNA sequences [24]. Our attempt to perform a census of the number of bacteria in a gram of soil provides a guidepost from which we can begin to assess the effects of environmental perturbations on community composition, diversity, evenness, and richness. Moreover, an accurate census would quantify the part of the microbial community that is not accounted for in the current models of community structure and function. In the Alaskan sequence collection, two sequences belonging to the sparsely sampled candidate phylum ACE were found only after sampling 832 sequences. We suspect that members of many poorly sampled candidate phyla are rare members in microbial communities [37], but may play significant functional roles in the microbial community. Although a reliable estimate of richness will inform the development of a conceptual framework for describing the functional biology of the soil microbial community, we will not know the texture and composition of that richness until we have exhaustively sampled and identified every member of the community.

MATERIALS AND METHODS

Clone Library Construction, Sequencing, and Analysis

We obtained a soil core from the Bonanza Creek Long-Term Ecological Research site approximately 30 km southwest of Fairbanks, Alaska, United States (64° 48′ N, 147° 52′ W) on the site designated BP-1 on an island in the Tanana River [38]. The L1A 16S rRNA gene library was constructed using a single 0.5-g sample of soil. The Bio101 soil DNA kit (Bio101, Irvine, California, United States) was used to extract and partially purify genomic DNA and the sample was further purified using a silica matrix (ExpressMatrix; Bio101) until it was suitable for PCR amplification.

16S rRNA genes were amplified in a single reaction by PCR using primers 27f (AGRGTTTGATYMTGGCTCAG) and 1492r (GGYTACCTTGTTACGACTT) and the products were purified by gel extraction (Qiaex II; Qiagen, Valencia, California, United States). Purified PCR products were ligated into the pGEM-T TA cloning vector as described by the manufacturer (Promega, Madison, Wisconsin, United States) and electroporated into E. coli(DH5α). Positive transformants were inoculated overnight into LB with ampicillin (100 μg/ml) and the culture was used as template for PCR using the universal M13f and M13r vector primers. These PCR products were purified using AmpPure (Agencourt Bioscience, Beverly, Massachusetts, United States) and sequenced using the 27f and 787r (CTACCRGGGTATCTAAT) primers. If the 787r primer did not produce quality sequence, we used either the M13f or the M13r primer for sequencing. Sequencing reactions were performed using BigDye version 3.1 (Applied Biosystems, Foster City, California,

United States) and were analyzed at the University of Wisconsin-Madison biotechnology center. All clones had 2-fold sequencing coverage for the first approximately 700 bp of the 16S rRNA gene.

Sequence contigs were constructed using STADEN [39] and aligned using ARB [40] with a reference database of more than 16,000 sequences longer than 1 kb. Putative chimeric sequences were identified using Bellerophon [41] and were further screened using CHIMERA_CHECK [42], partial treeing, and comparing the sequence alignment to predicted secondary structure to detect changes in helical base pairing and nucleotide signatures [43]. Phylogenetic placement of the 1,033 sequences was determined by identifying the phylum to which each sequence showed affinity after adding sequences to the database tree using the parsimony algorithm implemented in ARB with a 50% consensus mask.

We also obtained (from Susannah Green Tringe) two 16S rRNA gene sequence collections (N_{AKYG} = 875 sequences; N_{AKYH} = 758 sequences) constructed using a single 0.5-g sample of Minnesotan (Waseca County, Minnesota, United States [23]) farm soil. The original soil genomic DNA was obtained by cell fractionation followed by enzymatic and chemical extraction of the DNA [23,32]. Since our preliminary analysis using a nonparametric estimator of the fraction of shared OTUs [44] showed that the two Minnesota soil libraries harbored more than 68% of each others› $OTU_{0.03}$ membership, and they were made from the same soil sample but different PCR reactions, we pooled the 1,633 sequences into a single dataset. For direct comparison, the Minnesotan and Alaskan sequence collections were realigned using the NAST aligner [45] at the greengenes Web site (http://greengenes.lbl.gov), and the nucleotide sites between positions 150 and 700 (E. coli numbering) were used in subsequent analyses.

Community Analyses

To describe the community structure of each soil we used DOTUR's implementation of the furthest neighbor algorithm [18] to assign sequences to OTUs after exporting a Jukes-Cantor corrected distance matrix constructed in ARB using unmasked sequences. Output files from DOTUR were used to calculate collector's curves for the nonparametric estimators of the fraction of sequences from one library that affiliated with the OTUs shared between the libraries [44].

Model Community Analysis

Using a truncated lognormal, Pareto, or uniform distribution, we were able to construct model communities in which the probability of drawing an individual species followed a defined distribution. To identify the most appropriate truncated lognormal distribution that described the observed data, we first selected reasonable values for μ between 6.000 and 12.00 and S_T between 1,000 and 10,000. Next, using Equation 1, we identified values of σ that would yield N_T/N_{max} values of 10, 35, 40, 45, 50, 100, and 1,000. N_T/N_{max} is the reciprocal of the probability observed at the distribution's mode, where N_T represents the total number of individuals in a community and N_{max} represents the abundance of the most abundant member in the community. The values of N_T/N_{max} shown in Table 1 were selected because they fell within the range suggested by Curtis et al. [13] for microbial communities and because they resembled the frequency data observed from the Alaskan soil collection. For a given value of N_T/N_{max}, increasing μ increased the value of the reciprocal of the Simpson›s index (1/D). 1/Drepresents the number of uniformly abundant OTUs needed to observe the same level of diversity found in the community. Using these parameters, we drew random values from the desired lognormal distribution by first drawing a random normal variable with mean μ and standard deviation σ. Random lognormal variables were then obtained by determining the integer value of e^X, where X is the value of the random normal variable. Values larger than S_T were discarded, resulting in random variables drawn from a truncated lognormal distribution. Random variables drawn from either Pareto or uniformly distributed communities were done in an analogous manner.

As random values were generated, we constructed collector's curves for the observed richness and the full bias-corrected Chao1 [19], ACE [20], and interpolated Jackknife [21] nonparametric estimators. The heuristic search did not include the Jackknife estimator because the estimates were highly variable and uninformative. As measures of diversity, we determined 1/D and the longest string where duplicate members of the same taxa were not observed. We determined the CI for each metric as a function of sampling effort by constructing 1,000 model communities for each set of model parameters. The sampling of the truncated lognormal distributions and parameter calculation was performed using a C++ computer program that we wrote. If the collector›s curve for the sampling of the Alaskan or Minnesotan 16S rRNA sequence collections crossed the CI for any parameter, the simulated community was rejected.

ACKNOWLEDGMENTS

We appreciate the assistance of Susannah Green Tringe, who provided us with 16S rRNA gene sequences and details of the Minnesota soil microbial community characterization study.

AUTHOR CONTRIBUTIONS

PDS and JH conceived and designed the experiments. PDS performed and analyzed the experiments. PDS and JH wrote the paper.

REFERENCES

1. Whitman WB, Coleman DC, Wiebe WJ (1998) Prokaryotes: The unseen majority. Proc Natl Acad Sci USA 95: 6578–6583. WB WhitmanDC ColemanWJ Wiebe1998Prokaryotes: The unseen majority. Proc Natl Acad Sci USA9565786583
2. Begon M, Harper JL, Townsend CR (1996) Ecology: Individuals, populations, and communities. 3rd ed. Malden (Massachusetts): Blackwell Science. pp. 829–830. M. BegonJL HarperCR Townsend1996Ecology: Individuals, populations, and communities. 3rd edMalden (Massachusetts)Blackwell Sciencepp. 829830 pp.
3. Dunbar J, Takala S, Barns SM, Davis JA, Kuske CR (1999) Levels of bacterial community diversity in four arid soils compared by cultivation and 16S rRNA gene cloning. Appl Environ Microbiol 65: 1662–1669. J. DunbarS. TakalaSM BarnsJA DavisCR Kuske1999Levels of bacterial community diversity in four arid soils compared by cultivation and 16S rRNA gene cloning. Appl Environ Microbiol6516621669
4. Amann RI, Ludwig W, Schleifer KH (1995) Phylogenetic identification and in situ detection of individual microbial cells without cultivation. Microbiol Rev 59: 143–169. RI AmannW.

LudwigKH Schleifer1995Phylogenetic identification and in situ detection of individual microbial cells without cultivation. Microbiol Rev59143169

5. Ovreas L, Jensen S, Daae FL, Torsvik V (1998) Microbial community changes in a perturbed agricultural soil investigated by molecular and physiological approaches. Appl Environ Microbiol 64: 2739–2742. L. OvreasS. JensenFL DaaeV. Torsvik1998Microbial community changes in a perturbed agricultural soil investigated by molecular and physiological approaches. Appl Environ Microbiol6427392742
6. Ovreas L, Torsvik V (1998) Microbial diversity and community structure in two different agricultural soil communities. Microb Ecol 36: 303–315. L. OvreasV. Torsvik1998Microbial diversity and community structure in two different agricultural soil communities. Microb Ecol36303315
7. Torsvik V, Sorheim R, Goksoyr J (1996) Total bacterial diversity in soil and sediment communities—A review. J Indust Microbiol 17: 170–178. V. TorsvikR. SorheimJ. Goksoyr1996Total bacterial diversity in soil and sediment communities—A review. J Indust Microbiol17170178
8. Torsvik V, Goksoyr J, Daae FL (1990) High diversity in DNA of soil bacteria. Appl Environ Microbiol 56: 782–787. V. TorsvikJ. GoksoyrFL Daae1990High diversity in DNA of soil bacteria. Appl Environ Microbiol56782787
9. Dykhuizen DE (1998) Santa Rosalia revisited: Why are there so many species of bacteria? Anton Leeuw Int J G 73: 25–33. DE Dykhuizen1998Santa Rosalia revisited: Why are there so many species of bacteria?Anton Leeuw Int J G732533
10. Gans J, Wolinsky M, Dunbar J (2005) Computational improvements reveal great bacterial diversity and high metal toxicity in soil. Science 309: 1387–1390. J. GansM. WolinskyJ. Dunbar2005Computational improvements reveal great bacterial diversity and high metal toxicity in soil. Science30913871390
11. Sandaa R, Torsvik VV, Enger , Daae FL, Castberg T, et al. (1999) Analysis of bacterial communities in heavy metal-contaminated soils at different levels of resolution. FEMS Microbiol Ecol 30: 237–251. R. SandaaVV TorsvikEngerFL DaaeT. Castberg1999Analysis

of bacterial communities in heavy metal-contaminated soils at different levels of resolution. FEMS Microbiol Ecol30237251

12. Pace NR, Stahl DA, Lane DJ, Olsen GJ (1985) Analyzing natural microbial populations by rRNA sequences. ASM News 51: 4–12. NR PaceDA StahlDJ LaneGJ Olsen1985Analyzing natural microbial populations by rRNA sequences. ASM News51412
13. Curtis TP, Sloan WT, Scannell JW (2002) Estimating prokaryotic diversity and its limits. Proc Natl Acad Sci USA 99: 10494–10499. TP CurtisWT SloanJW Scannell2002Estimating prokaryotic diversity and its limits. Proc Natl Acad Sci USA991049410499
14. Lunn M, Sloan WT, Curtis TP (2004) Estimating bacterial diversity from clone libraries with flat rank abundance distributions. Environ Microbiol 6: 1081–1085. M. LunnWT SloanTP Curtis2004Estimating bacterial diversity from clone libraries with flat rank abundance distributions. Environ Microbiol610811085
15. Hughes JB, Hellmann JJ, Ricketts TH, Bohannan BJM (2001) Counting the uncountable: Statistical approaches to estimating microbial diversity. Appl Environ Microbiol 67: 4399–4406. JB HughesJJ HellmannTH RickettsBJM Bohannan2001Counting the uncountable: Statistical approaches to estimating microbial diversity. Appl Environ Microbiol6743994406
16. Hong SH, Bunge J, Jeon SO, Epstein SS (2006) Predicting microbial species richness. Proc Natl Acad Sci USA 103: 117–122. SH HongJ. BungeSO JeonSS Epstein2006Predicting microbial species richness. Proc Natl Acad Sci USA103117122
17. Dunbar J, Barns SM, Ticknor LO, Kuske CR (2002) Empirical and theoretical bacterial diversity in four Arizona soils. Appl Environ Microbiol 68: 3035–3045. J. DunbarSM BarnsLO TicknorCR Kuske2002Empirical and theoretical bacterial diversity in four Arizona soils. Appl Environ Microbiol6830353045
18. Schloss PD, Handelsman J (2005) Introducing DOTUR, a computer program for defining operational taxonomic units and estimating species richness. Appl Environ Microbiol 71: 1501–1506. PD SchlossJ. Handelsman2005Introducing DOTUR, a computer program for defining operational taxonomic units and estimating species richness. Appl Environ Microbiol7115011506

19. Chao A (1984) Non-parametric estimation of the number of classes in a population. Scand J Stat 11: 265–270. A. Chao1984Non-parametric estimation of the number of classes in a population. Scand J Stat11265270

20. Chao A, Ma MC, Yang MCK (1993) Stopping rules and estimation for recapture debugging with unequal failure rates. Biometrika 80: 193–201. A. ChaoMC MaMCK Yang1993Stopping rules and estimation for recapture debugging with unequal failure rates. Biometrika80193201

21. Burnham KP, Overton WS (1979) Robust estimation of population size when capture probabilities vary among animals. Ecology 60: 927–936. KP BurnhamWS Overton1979Robust estimation of population size when capture probabilities vary among animals. Ecology60927936

22. McCaig AE, Glover LA, Prosser JI (1999) Molecular analysis of bacterial community structure and diversity in unimproved and improved upland grass pastures. Appl Environ Microbiol 65: 1721–1730. AE McCaigLA GloverJI Prosser1999Molecular analysis of bacterial community structure and diversity in unimproved and improved upland grass pastures. Appl Environ Microbiol6517211730

23. Tringe SG, von Mering C, Kobayashi A, Salamov AA, Chen K, et al. (2005) Comparative metagenomics of microbial communities. Science 308: 554–557. SG TringeC. von MeringA. KobayashiAA SalamovK. Chen2005Comparative metagenomics of microbial communities. Science308554557

24. Keswani J, Whitman WB (2001) Relationship of 16S rRNA sequence similarity to DNA hybridization in prokaryotes. Int J Syst Evol Microbiol 51: 667–678. J. KeswaniWB Whitman2001Relationship of 16S rRNA sequence similarity to DNA hybridization in prokaryotes. Int J Syst Evol Microbiol51667678

25. Sait M, Hugenholtz P, Janssen PH (2002) Cultivation of globally distributed soil bacteria from phylogenetic lineages previously only detected in cultivation-independent surveys. Environ Microbiol 4: 654–666. M. SaitP. HugenholtzPH Janssen2002Cultivation of globally distributed soil bacteria from phylogenetic lineages previously only detected in cultivation-independent surveys.

Environ Microbiol4654666

26. Stackebrandt E, Goebel BM (1994) A place for DNA-DNA reassociation and 16S rRNA sequence-analysis in the present species definition in bacteriology. Int J Syst Bacteriol 44: 846–849. E. StackebrandtBM Goebel1994A place for DNA-DNA reassociation and 16S rRNA sequence-analysis in the present species definition in bacteriology. Int J Syst Bacteriol44846849
27. Everett KDE, Bush RM, Andersen AA (1999) Emended description of the order Chlamydiales, proposal of Parachlamydiaceae fam. nov. and Simkaniaceae fam. nov. , each containing one monotypic genus, revised taxonomy of the family Chlamydiaceae,including a new genus and five new species, and standards for the identification of organisms. Int J Syst Bacteriol 49: 415–440. KDE EverettRM BushAA Andersen1999Emended description of the order Chlamydiales, proposal of Parachlamydiaceae fam. nov. and Simkaniaceae fam. nov. , each containing one monotypic genus, revised taxonomy of the family Chlamydiaceae, including a new genus and five new species, and standards for the identification of organisms. Int J Syst Bacteriol49415440
28. Hugenholtz P, Goebel BM, Pace NR (1998) Impact of culture-independent studies on the emerging phylogenetic view of bacterial diversity. J Bacteriol 180: 4765–4774. P. HugenholtzBM GoebelNR Pace1998Impact of culture-independent studies on the emerging phylogenetic view of bacterial diversity. J Bacteriol18047654774
29. Klappenbach JA, Saxman PR, Cole JR, Schmidt TM (2001) RMDB: The ribosomal RNA operon copy number database. Nucleic Acids Res 29: 181–184. JA KlappenbachPR SaxmanJR ColeTM Schmidt2001RMDB: The ribosomal RNA operon copy number database. Nucleic Acids Res29181184
30. Clayton RA, Sutton G, Hinkle PS, Bult C, Fields C (1995) Intraspecific variation in small-subunit rRNA sequences in Genbank—Why single sequences may not adequately represent prokaryotic taxa. Int J Syst Bacteriol 45: 595–599. RA ClaytonG. SuttonPS HinkleC. BultC. Fields1995Intraspecific variation in small-subunit rRNA sequences in Genbank—Why single sequences may not adequately represent prokaryotic taxa. Int J Syst Bacteriol45595599

31. Magurran AE (2004) Measuring biological diversity. Malden (Massachusetts): Blackwell. 256 p. AE Magurran2004Measuring biological diversityMalden (Massachusetts)Blackwell256p.

32. Robertson DE, Chaplin JA, DeSantis G, Podar M, Madden M, et al. (2004) Exploring nitrilase sequence space for enantioselective catalysis. Appl Environ Microbiol 70: 2429–2436. DE RobertsonJA ChaplinG. DeSantisM. PodarM. Madden2004Exploring nitrilase sequence space for enantioselective catalysis. Appl Environ Microbiol7024292436

33. Miller DN, Bryant JE, Madsen EL, Ghiorse WC (1999) Evaluation and optimization of DNA extraction and purification procedures for soil and sediment samples. Appl Environ Microbiol 65: 4715–4724. DN MillerJE BryantEL MadsenWC Ghiorse1999Evaluation and optimization of DNA extraction and purification procedures for soil and sediment samples. Appl Environ Microbiol6547154724

34. Waksman SA, Starkey RL (1931) The Soil and the microbe. New York: John Wiley. 260 p. SA WaksmanRL Starkey1931The Soil and the microbeNew YorkJohn Wiley260p.

35. Wilson EO (1999) The diversity of life. New York: W. W. Norton. 424 p. EO Wilson1999The diversity of lifeNew YorkW. W. Norton424p.

36. Riesenfeld CS, Schloss PD, Handelsman J (2004) Metagenomics: Genomic analysis of microbial communities. Annu Rev Genet 38: 525–552. CS RiesenfeldPD SchlossJ. Handelsman2004Metagenomics: Genomic analysis of microbial communities. Annu Rev Genet38525552

37. Schloss PD, Handelsman J (2004) Status of the microbial census. Microbiol Mol Biol Rev 68: 686–691. PD SchlossJ. Handelsman2004Status of the microbial census. Microbiol Mol Biol Rev68686691

38. Williamson LL, Borlee BR, Schloss PD, Guan C, Allen HK, et al. (2005) Intracellular screen to identify metagenomic clones that induce or inhibit a quorum-sensing biosensor. Appl Environ Microbiol 71: 6335–6344. LL WilliamsonBR BorleePD SchlossC. GuanHK Allen2005Intracellular screen to identify metagenomic clones that induce or inhibit a quorum-sensing biosensor. Appl

Environ Microbiol7163356344

39. Bonfield JK, Smith KF, Staden R (1995) A new DNA sequence assembly program. Nucleic Acids Res 23: 4992–4999. JK BonfieldKF SmithR. Staden1995A new DNA sequence assembly program. Nucleic Acids Res2349924999
40. Ludwig W, Strunk O, Westram R, Richter L, Meier H, et al. (2004) ARB: A software environment for sequence data. Nucleic Acids Res 32: 1363–1371. W. LudwigO. StrunkR. WestramL. RichterH. Meier2004ARB: A software environment for sequence data. Nucleic Acids Res3213631371
41. Huber T, Faulkner G, Hugenholtz P (2004) Bellerophon: A program to detect chimeric sequences in multiple sequence alignments. Bioinformatics 20: 2317–2319. T. HuberG. FaulknerP. Hugenholtz2004Bellerophon:Aprogramtodetectchimericsequences in multiple sequence alignments. Bioinformatics2023172319
42. Cole JR, Chai B, Marsh TL, Farris RJ, Wang Q, et al. (2003) The ribosomal database project (RDP-II): Previewing a new autoaligner that allows regular updates and the new prokaryotic taxonomy. Nucleic Acids Res 31: 442–443. JR ColeB. ChaiTL MarshRJ FarrisQ. Wang2003The ribosomal database project (RDP-II): Previewing a new autoaligner that allows regular updates and the new prokaryotic taxonomy. Nucleic Acids Res31442443
43. Hugenholtz P, Huber T (2003) Chimeric 16S rDNA sequences of diverse origin are accumulating in the public databases. Int J Syst Evol Micr 53: 289–293. P. HugenholtzT. Huber2003Chimeric 16S rDNA sequences of diverse origin are accumulating in the public databases. Int J Syst Evol Micr53289293
44. Chao A, Chazdon RL, Colwell RK, Shen TJ (2005) A new statistical approach for assessing similarity of species composition with incidence and abundance data. Ecol Lett 8: 148–159. A. ChaoRL ChazdonRK ColwellTJ Shen2005A new statistical approach for assessing similarity of species composition with incidence and abundance data. Ecol Lett8148159
45. DeSantis TZ, Dubosarskiy I, Murray SR, Andersen GL (2003) Comprehensive aligned sequence construction for automated design of effective probes (CASCADE-P) using 16S rDNA. Bioinformatics

19: 1461–1468. TZ DeSantisI. DubosarskiySR MurrayGL Andersen2003Comprehensive aligned sequence construction for automated design of effective probes (CASCADE-P) using 16S rDNA. Bioinformatics1914611468

8

CHAPTER

VARIATIONS OF BACTERIAL COMMUNITY STRUCTURE AND COMPOSITION IN MANGROVE SEDIMENT AT DIFFERENT DEPTHS IN SOUTHEASTERN BRAZIL

Lucas William Mendes and Siu Mui Tsai

Cell and Molecular Biology Laboratory, Center for Nuclear Energy in Agriculture CENA, University of São Paulo USP, Piracicaba, SP, CEP 13400-970, Brazil

ABSTRACT

Tropical mangroves are considered one of the most productive ecosystems of the world, being characterized as nurseries and food sources for fish and other animals. Microorganisms play important roles in these environments, and the study of bacterial communities is of paramount importance for a better comprehension of mangrove dynamics. This study focused on the structure and composition of bacterial communities in mangrove sediments at different depths and points, located in Southeastern Brazil. Terminal Restriction Fragment Length Polymorphism (T-RFLP) was used to determine the community structure, and 16S rRNA gene pyrosequencing was used to characterize the community composition. Redundancy analysis of T-RFLP patterns revealed differences in bacterial community structure according to soil attributes and depth. The parameters K and depth presented significant correlation with general community structure. Most sequences were classified into the phylum *Proteobacteria* (88%), which presented differences according to the depth, where the classes *Betaproteobacteria* (21%) and *Deltaproteobacteria* (16%) were abundant at 10 cm and *Epsilonproteobacteria* (35%) was abundant at 40 cm depth. Clear differences were observed in community composition as shown by the differential distribution of the phyla *Firmicutes* (1.13% and 3.8%, for 10 cm and 40 cm respectively), *Chloroflexi* (2.8% and 0.75%), and *Acidobacteria* (2.75% and 0.57%) according to the depth. Bacterial diversity measurements indicated higher diversity in shallow samples. Taken together, our findings indicate that mangrove holds a diverse bacterial community, which is shaped by the variations found in the ecosystem, such as sediment properties and depth.

Keywords:

microbial ecology; T-RFLP; pyrosequencing; 16S rRNA; tropical mangrove; sediment bacterial community

INTRODUCTION

Tropical mangroves are among the most productive ecosystems in the world, being characterized by high rate of organic matter and nutrient cycling [1]. They are located in the intertidal zones along tropical and subtropical coasts, in the transition between terrestrial, marine and freshwater environments [2]. Mangroves constitute a complex environment with high interaction between plants, animals and microorganisms, with important features such as anaerobic and highly reducing sediment [2,3]. The main determining factors for the occurrence, structure and functioning of these ecosystems are tropical temperatures, alluvial substrates, protection against waves, presence of salt water and large tidal range [4]. They are important breeding and growth habitats, refuge and food source for fish and other animals [5]. Although their importance is well known, they are at risk of degradation by pollutants and deforestation, needing urgent attention and conservation programs [6]. In mangrove sediment, the microorganisms play an important role in the productivity, conservation, and environment recovery, where they participate in the biogeochemical cycles and supply plant and animals with primary nutritional sources [7,8]. The structure and composition of microbial communities in mangrove are influenced by several factors such as nutrient content, biogeography, depth, anthropogenic activities, and tidal regime [8,9,10,11]. In recent years, the microbial communities of mangroves have been assessed using a range of techniques, including classical cultivation approach, fingerprinting techniques, clone libraries and next generation sequencing [8,9,11,12]. These studies focused on the diversity and distribution of microbial communities at different sites [8,10], different points within a mangrove [8], state of preservation [9,11,13], contamination by oil spill [14] and shrimp farms [15], and depths [11]. Recently, the advance in the next-generation DNA sequence methods, such as pyrosequencing [16], boosts scientific interests to understand the complexity of microbial community in a wide range of environments.

Considering the importance and complexity of mangrove ecosystems, a description of the distribution and composition of bacterial communities in sediments would lead us to a better comprehension of the ecosystem dynamics. In the present study, in order to obtain a better insight into the mangrove microbiome, we assessed the bacterial community structure (T-RFLP) and composition (pyrosequencing) in a mangrove at different points and depths, and correlated this structure with the sediment chemical properties. This work highlights the importance of microbial ecology studies in mangrove environments and provides information that can be suitable for future development of sustainable approaches for the ecosystem preservation.

METHODS

Sampling

Sediment samples were collected from a pristine mangrove at Cardoso Island State Park in the estuary of the Cananéia, São Paulo State, Southeastern Brazil (25°05′02″S and 47°57′42″W—Point 1) in summer

2007. The vegetation is composed of three species, *Laguncularia racemosa*, *Avicennia shaueriana* and *Rhizophora mangle*. Sediment samples were collected along a transect of 180° S orientation and began at the margin of the bay and extended to the land. Three sub-regions were defined based on their proximity to the sea (Point 1), the middle of the mangrove (Point 2) and the proximity to the land (Point 3). The points were distributed equidistantly and were separated from one another by approximately 70 meters. Each point samples were collected in triplicate, avoiding the rhizospheric effect caused by the presence of plants roots on sediments. Tubes of 60 cm in length and 10 cm in diameter were inserted into the sediment and recovered to obtain sediment cores, which provided a vertical profile of the sediments. The cores were kept in the vertical orientation and on ice until their further processing in lab. Each of the cores was separated into subsamples based on sediment depth. Subsamples were collected at four depths: 0–10 cm, 10–20 cm, 20–30 cm and 30–40 cm. Samples were taken from the center of the tube to avoid any contamination. For each subsample, 1.5 g of sediment was collected and kept −80 °C until DNA extraction.

Sediment Chemical and Physical Analysis

Sediment chemical properties were determined in triplicate for each sampling point and depth, based on 400 g of soil, performed at the Laboratory of Soil Analysis at "Luiz de Queiroz" College of Agriculture (ESALQ/USP, Piracicaba, Brazil), according to the methodology described by Van Raij *et al.*,. [17]. Briefly, the soil pH was measured in a 1:2.5 soil/water suspension and Organic matter (OM) content was measured via colorimetric method. Exchangeable Ca, and Mg were extracted with KCL 1 M. Calcium and Mg were determined by atomic absorption. Phosphorus and K were extracted by ion-exchange resin. Available micronutrients (Fe, Mn, Zn, and Cu) were extracted by Mehlich 1 and determined by atomic absorption spectrometry. Boron was extracted with hot water and determined by spectrophotometry with azomethine-H at 420 nm. Soil organic C was determined by the Walkley-Black method. The content of sand, silt and clay was estimated by dry soil and other procedures were made according to the methodologies used routinely in the lab. Nutrient content, pH and OM were measured according to the depth for each point, and physical properties, carbon, nitrogen, and sulfur were measured according to the point of sampling regardless of the depth.

DNA Extraction and T-RFLP Analysis

DNA extraction from 250 mg of soil sample was carried out using the PowerSoil DNA Isolation Kit (MoBio Laboratories, Carlsbad, CA, USA), according to the manufacturer's protocol. DNA quality and concentration were measured using NanoDrop 1000 spectrophotometry (Thermo Scientific, Waltham, EUA) and 1% sodium boric acid [18] agarose gel electrophoresis. T-RFLP fingerprinting was used to characterize the bacterial community structure of mangrove sediments. The 16S rRNA gene was amplified with primers 27f (5' AGAGTTTGATCCTGGCTCAG 3') e 1492r (5' ACCTTGTTACGACTT 3') [19]. The forward primer was labeled with 6-FAM (6-carboxyfluorescein) at the 5' end. The 25 μL reaction mixture contained 2.5 μL reaction buffer 10X (Invitrogen, Carlsbad, CA, USA), 1.0 μL $MgCl_2$ (50 mM), 0.5 μL of deoxyribonucleotide triphosphate mixture (2.5 mM each), 0.25 μL of bovine serum albumin (BSA) (0.5 $\mu g \cdot \mu L^{-1}$), 0.5 μL of the primer set (5 pmol each), 0.2 μL (5 U) of *Platinum Taq* DNA polymerase (Invitrogen), 5 μL of DNA template (approximately 50 ng) and 14.55 μL of sterilized ultrapure water. PCR amplification was performed using a GeneAmp PCR System 9700 thermal cycler (Applied Biosystems, Foster City, CA, USA) with the following amplification cycles: 94 °C for 3 min (initial denaturation), followed by 35 cycles of 94 °C for 30 s, 59 °C for 45 s, and 72 °C for 1 min with a final extension step at 72 °C for 15 min. Negative controls without DNA were run in all amplifications. PCR products were purified using the GFX™ PCR DNA and Gel Band purification kit (GE Healthcare, Little Chalfont, Buckinghamshire, UK) after analysis by gel electrophoresis. After purification, 5 μL of amplification product (nearly 60 ng) were digested with the endonuclease *HhaI* (Invitrogen) in 15 μL reactions for 3 h at 37 °C. The digests were purified by sodium acetate/EDTA precipitation and then mixed with 0.25 μL of the Genescan 500 ROX size standard (Applied Biosystems) and 9.75 μL of deionized formamide. Prior to fragment analysis, samples were denatured at 95 °C for 5 min and chilled on ice. Analysis of terminal restriction fragment sizes and quantities was performed on an ABI PRISM 3100 genetic analyzer (Applied Biosystems).

T-RFLP profiles were analyzed using PeakScanner v1.0 software (Applied Biosystems, Foster City, CA, USA). Terminal restriction fragments (T-RFs) smaller than 50 bp and larger than 800 bp were excluded from the analysis to avoid detection of primers and uncertainties of size determination. True peaks were determined using T-REX online software [20]. T-RF sizes were rounded to the nearest integer and peak heights were normalized to account for uncontrolled differences in the quantity of DNA between samples. The relative abundance of a single T-RF was calculated as percent fluorescence intensity relative to total fluorescence intensity of the peaks [21]. Diversity measurements (richness and Shannon index) were calculated based on T-RF abundance matrix and statistical analyzes of data were performed using One-way ANOVA based on Tukey's test to determine the significance of differences between samples, using the software PAST [22]. Data from individual samples were subjected to multivariate analysis using the software Canoco 4.5 (Biometris, Wageningen, The Netherlands) and Primer6 (PrimerE, Ivybridge, UK) to determine the correlation between community structure and sediment physicochemical properties. All matrices were initially analyzed using de-trended correspondence analysis (DCA) to evaluate the gradient size of the species distribution; this analysis indicated linearly distributed data (length of gradient < 3), revealing that the best-fit mathematical model for the data was redundancy analysis (RDA). Forward selection (FS) and the Monte Carlo permutation test were applied with 1000 random permutations to verify the significance of sediment physicochemical properties upon a microbial community. In addition to *p* values for the significance of each soil chemical property, RDA and Monte Carlo permutation tests supplied information about the marginal effects of environmental variables, quantifying the amount of variance explained by each factor.

Pyrosequencing Analysis of the Bacterial 16S rRNA Gene

To assess the composition of bacterial communities in mangrove sediments, samples from 10 cm and 40 cm depths were submitted to pyrosequencing analysis. The samples from points 1 to 3 were pooled and we consider for sequencing only the different depths of 10 cm and 40 cm. A partial V4 region of the bacterial 16S rRNA gene was amplified using the primer set 5' AYTGGGYDTAAAGNG 3' and 5' TACCRGGGTHTCTAATCC 3' containing specific Roche 454-pyrosequencing adaptors and barcodes of 8 bp [23]. PCRs contained 1X reaction buffer, 1.8 mM·$MgCl_2$, 0.2 mM of each dNTP, 0.2 μM of each primer (Integrate DNA Technology), 10 mg·mL^{-1} of BSA, 4 ng·μL^{-1} of template, and 1 unit of FastStart High Fidelity PCR system enzyme blend (Roche Applied Science, Indianapolis, IN, USA) adding to 50 μL. Amplifications were performed using the following cycle parameters: 95 °C for 3 min (initial denaturation), followed by 30 cycles of 95 °C for 40 s, 57 °C for 45 s, and 72 °C for 1 min with a final extension step at 72 °C for 4 min. The PCR products were separated by gel electrophoresis, and fragments with size in the range of 270–300 bp were excised from the gel and extracted by using the Qiagen Gel Extraction Kit (Qiagen, Valencia, CA, USA). Further purification was performed with Qiagen PCR Purification Kit (Qiagen, Valencia, CA, USA). Sequencing was performed using a GS FLX sequencer (454 Life Sciences) at the Michigan State University Research Technology Support Facility using Titanium chemistry.

Raw sequences were processed through the Ribosomal Database Project (RDP) pyrosequencing pipeline (http://pyro.cme.msu.edu). Within this tool the following analysis was performed, base calling, barcode sorting, primer trimming, chimera detection, and removal of bad quality sequences. The sequences obtained as described above were checked for presence of chimeras using the Greengenes chimera-check tool [24]. The remaining sequences were then analyzed using the Mothur software [25] based on a cut-off of 97% similarity, in order to identify operational taxonomic units (OTUs), measure diversity indices, generate rarefaction curves and Venn diagrams and to test significant difference between samples using the Libshuff test [26]. The sequence data generated by 454-pyrosequencing are available on the MG-RAST server (http://metagenomics.anl.gov) under the project "Bacterial Mangrove Profile" (ID 10820) and accession number 4481551.3 and 4481552.3 for 10 and 40 cm depths, respectively.

RESULTS AND DISCUSSION

The aim of this study was to assess the bacterial community structure and composition in sediments of Brazilian mangrove at different points and depths. Previous studies have shown the high microbial diversity in these environments, indicating that they might play an important role in nutrient cycling [8,9,10,11,14,27]. However, considering that mangrove ecosystems are very dynamic, in this study we took into account different points and depths. The sediment samples collected at the mangrove site were subjected to physical and chemical analyzes, and some factors were different according to the point or depth. The Table 1A shows the result of pH, OM and nutrients analyzed according to the depth for each sampling point. The Table 1B shows the result of carbon, nitrogen and sulfur content, and physical characteristics measured only for the different points in the mangrove, regardless the depth. The overall trend for the observed parameters was decrease over the vertical profile. Regarding to the points of sampling, the soil texture, carbon, conductivity and salinity were different between points. Environmental factors as salinity and nutrient availability are very variable and influenced by periodic tidal flood [28,29]. Bacterial communities are very responsive to this constant change in environmental conditions in order to adapt and maintain the ecosystem functioning [29]. Our data point to the importance of considering differences in mangrove ecosystem in microbial ecology studies.

Table 1: Physicochemical properties of mangrove sediments sampled at different (A) depths and (B) points

	Point 1				Point 2				Point 3			
A	**10 cm**	**20 cm**	**30 cm**	**40 cm**	**10 cm**	**20 cm**	**30 cm**	**40 cm**	**10 cm**	**20 cm**	**30 cm**	**40 cm**
pH ($CaCl_2$)	6.1	6.6	6.6	5.7	6.3	6.7	6.5	5.7	5.4	6.4	6.1	6.2
OM ($g \cdot dm^{-3}$)	94	124	88	40	122	91	100	54	51	98	109	46
P ($mg \cdot dm^{-3}$)	12	8	7	3	6	8	7	4	2	6	4	3
K ($mmol \cdot dm^{-3}$)	2.3	1.7	1.7	1.1	2.6	2.1	1.8	1.4	1.1	1.8	1.8	1.4
Ca ($mmol \cdot dm^{-3}$)	49	47	43	21	45	49	29	18	16	37	28	19
Mg ($mmol \cdot dm^{-3}$)	55	55	51	38	18	15	15	18	20	15	18	13
B ($mg \cdot dm^{-3}$)	0.1	0.05	0.04	0.02	0.1	0.06	0.05	0.01	0.12	0.05	0.05	0.02
Cu ($mg \cdot dm^{-3}$)	0.2	0.07	0.06	0.02	0.19	0.08	0.06	0.02	0.2	0.07	0.06	0.01
Fe ($mg \cdot dm^{-3}$)	281	59.1	55.1	53	260	53.2	40.1	47.8	280	70.3	59.8	54
Mn ($mg \cdot dm^{-3}$)	2.92	1.08	0.96	0.96	1.98	1.00	0.80	0.91	2.40	1.10	0.85	0.70
Zn ($mg \cdot dm^{-3}$)	0.6	0.05	0.06	0.06	0.6	0.06	0.06	0.05	0.5	0.05	0.04	0.05

B	**Point 1**	**Point 2**	**Point 3**
Sand (%)	22	82	78
Silt (%)	26	6	12
Clay (%)	52	12	10
Temperature (°C)	12.5	13.3	13.7
Humidity (%)	69.9	40.8	49.2
Total carbon (%)	13.9	5.1	7.0
Organic carbon (%)	12.2	3.0	3.5
Total nitrogen (%)	0.7	0.2	0.2
Total sulphur (%)	0.2	0.1	0.1
Condutivity (mS)	20.8	11.4	3.5
Salinity (%)	1.3	0.7	0.2

According to T-RFLP patterns, the bacterial community structure was different among samples, and the difference was larger between depths than points. Redundancy analysis showed that the observed variation between community structure was correlated with sediment attributes and environmental variations, such as conductivity, pH, point, potassium content and depth, the last two being the most significant (Figure 1 and Table 2). Several studies support that pH and attributes related to soil acidity are the best predictors of the richness and diversity of microbial communities in soils [30,31,32,33]. In our work, the pH showed significant difference between 10 and 40 cm depth, which may contribute to the difference found in bacterial structure. The K was the nutrient that most correlated to the variation in the community structure. Soil microorganisms have been reported to play a key role in the K cycle, and the potassium solubilizing bacteria are important to make potassium available for uptake by plants [34]. One of the best-known groups of potassium solubilizing bacteria is *Bacillus*, which belongs to the class *Bacilli* of the phylum *Firmicutes* [35,36]. Interestingly, our samples presented a clear variation in the abundance of sequences of the class *Bacilli* according to the depth. From the total sequences of *Firmicutes*, 2% and 47% belong to the class *Bacilli* for the samples from 10 cm and 40 cm depths, respectively. Interestingly, according to the chemical analysis, the sample from 40 cm presented low amount of K. Studies on long-term fertilization showed that microbial community responds to alteration in K content [37,38]. Together with K, the factor "depth" is significantly correlated to the bacterial community structure ($p < 0.05$). Although the Shannon diversity did not presented difference ($p > 0.05$), the richness was significantly different between 10 and 40 cm depth ($p < 0.05$), with shallow samples presenting higher number of T-RFs (Figure 2). In a previous study, Mendes *et al.,*. [11] showed that archaeal communities in mangrove are different at different depths, and the shallow samples presented higher diversity. In a study with methanogenic *Archaea* in mangrove, Taketani *et al.,*. [14] show a decrease in the richness of species belonging to the phylum *Euryarchaeota* with increasing depth. In both studies cited above, they correlate these differences to the decrease of OM according to the depth. In our study, other factors than OM also decreased with the increased depth, such as K, Mg and Fe (Table 1). This variation of nutrients may explain the significant difference between 10 and 40 cm depths. Along with our findings, several studies have also demonstrated that the microbial structure responds to different depths in soils [39,40,41,42].

Table 2: Statistical significance of environmental variable determined by multivariate analysis based on the redundancy methodology (RDA)

Variable	Lambdal	*p* Value
K	**0.25**	**0.044**
Depth	**0.22**	**0.035**
Conductivity	0.16	0.064
Point	0.03	0.474
pH	0.01	0.967

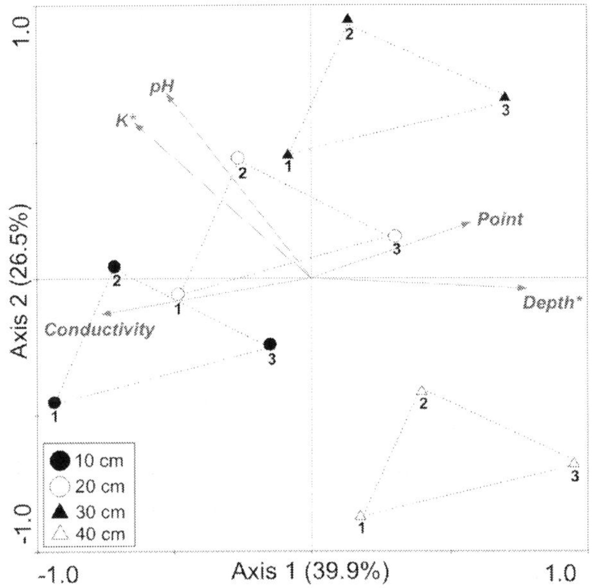

Figure 1: Redundancy analysis (RDA) performed on T-RFLP patterns obtained for the bacterial 16S rRNA gene and environmental characteristics of the mangrove sediment. Arrows indicate correlation between the chemical parameters and sample community structure. The significance of these correlations was evaluated by the Monte Carlo permutation test and is indicated as follows: * $p < 0.05$ (see also Table 2). The number below the symbols indicates the point of sampling.

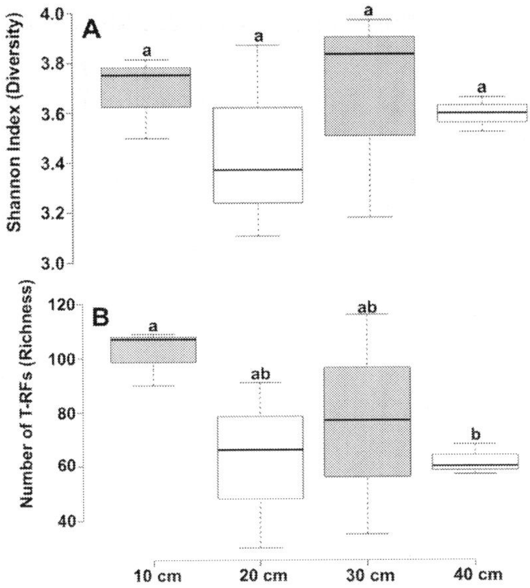

Figure 2: Boxplots of the Shannon's index and richness of observed T-RFs for mangrove bacterial communities in four depths. Center lines show the medians; box limits indicates the 25th and 75th percentiles as determined by R software; whiskers extend 1.5 times the interquartile range from the 25th and 75th percentiles. Significant differences among sediment depths are indicated by different letters according to the performed ANOVA and Tukey HSD test ($p < 0.05$).

The T-RFLP analysis indicated significant variation of bacterial community according to depth, therefor to further investigate the bacterial composition in mangrove sediment at different depths, we applied 454 pyrosequencing of the V4 hypervariable region of the 16S rRNA gene for samples from 10 cm and 40 cm depths. In this study, the description of bacterial communities using a culture-independent approach enable an overview of the groups inhabiting each depth of mangrove. The number of sequence reads obtained for each sampling was 6382 and 7117 sequences for 10 cm and 40 cm depths, respectively (Table 3). Statistical analyses were applied to evaluate species richness, the number of OTUs, and the difference occurring between the communities from the two different depths. The OTU were defined at a cut-off of 97% sequence similarity, a commonly recognized level for comparative analysis of 16S rRNA gene [43]. Confirming the results provided by T-RFLP analysis, differences between depths within the mangrove were detected by the richness of species and Chao1 richness estimator, where shallow samples presented higher number of OTUs (Table 3). In addition, the 454 sequencing showed that 10 cm depth presented higher diversity as shown by Shannon index (Table 3). This result is supported by the rarefaction curves (Figure 3), which indicated that 40 cm sampling requires fewer sequences to ensure comprehensive sampling effort at this depth compared to 10 cm sample. This rarefaction trend is an indicative of the diversity present in shallow layer of mangroves, higher than in the deeper layers. Despite the rarefaction curves have shown differences between depths, the curve did not achieve the plateau, revealed that this result should be viewed as an underestimation of the total diversity due the amount of sequences sampled [44]. Figure 4shows the Venn diagram based on the 16S rRNA sequences from both mangrove depths. The sequences were separated into bins based on 0.03 cut-off dissimilarity, using Mothur to highlight the number of OTUs shared between samples. The number of shared sequences between depths was <16%, which reveals difference in the bacterial composition between 10 cm and 40 cm. Our findings are in accordance of others studies that found a low percentage of sequences shared between shallow and deeper samples [11,14]. This difference is confirmed by the significant difference pointed by the Libshuff test ($p <$ 0.001).

Table 3: Comparison of diversity estimation of 16S rRNA sequencing from mangrove samples

Depth	Reads	OTU [a]	Chao1 [b]	Shannon	Good
10 cm	6382	1580	2840.65	5.97	81.1
40 cm	7117	1234	2285.88	5.62	79.0

[a] The operational taxonomic units (OTUs) were defined with 3% dissimilarity; [b] The richness estimator (Chao1), diversity index (Shannon) and coverage percentage (Good 1953) were calculated using Mothur (Schloss *et al.*, 2009).

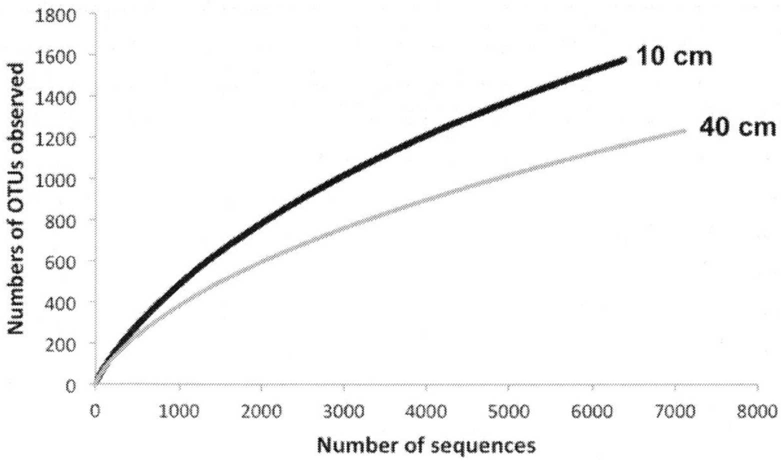

Figure 3: Rarefaction curves generated for 16S rRNA pyrosequence from samples collected at 10 and 40 cm depth in mangrove sediment. Sequences were grouped into OTUs based on a distance sequence similarity of 97%.

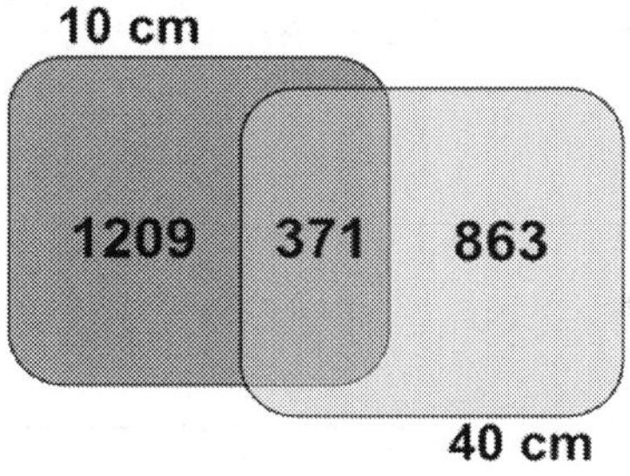

Figure 4: Venn diagram of sequences from mangrove samples at 10 and 40 cm depths. Sequences were grouped into OTUs based on distance sequence similarity of 97% separated by Mothur.

The obtained sequences were classified using the RDP database, and the dominant phylum in all samples was *Proteobacteria*, with 88% of total sequences (Figure 5). The phyla that presented the highest differences between depths were *Firmicutes* (1.13% of sequences for 10 cm and 3.8% for 40 cm), *Chloroflexi* (2.8% for 10 cm and 0.75% for 40 cm) and *Acidobacteria* (2.73% for 10 cm and 0.57% for 40 cm). Although in different proportions, our results are similar to others that used 16S rRNA to elucidate the bacterial diversity in mangrove ecosystems

[8,9,13,14,45]. Regarding the sequences shared between the depths, half of them were unclassified into Bacteria Domain (50.2%) and 34% were classified as *Proteobacteria.* Within this phylum, most of sequences shared between depths were classified as *Gammaproteobacteria*(28.3%) and *Alphaproteobacteria* (21.6%) classes. Within the phylum *Proteobacteria,* the classes *Betaproteobacteria* (21%) and *Deltaproteobacteria* (16%) were abundant at 10 cm, while *Epsilonproteobacteria* (35%) was abundant at 40 cm depth. *Betaproteobacteria* plays a role in nitrogen fixation [46] and several groups are aerobic or facultative, which explain higher abundance at shallow samples. Members belonging to the class *Deltaproteobacteria* participate in iron and sulfate reduction [47,48]. In our samples, there was a decrease of sequences belonging to *Deltaproteobacteria* from 10 cm (16%) to 40 cm (4%). In shallow samples, this phylum was dominated by the order *Desulfobacterales* (52.5%), while in deeper samples were dominated by *Desulfobacterales* (23.3%) and *Desulfuromonadales* (13.6%). This result is similar to the work of Dos Santos *et al.,.* [49], which also found a dominance of *Desulfobacterales* that is mostly composed of anaerobic microorganisms and involved in sulfur cycling. Andreote *et al.,.* [8] showed that members of *Desulfobacterales* are also involved in the transformation of methane and nitrogen, which make this group one of the most important in mangrove ecosystem. The *Epsilonproteobacteria* are commonly found in deep-sea hydrothermal vents, and are characterized by chemolithotrophy, meeting their energy needs by oxidizing reduced sulfur [50]. One of the most important characteristics of mangrove sediments is the absence of oxygen at a few millimeters below the surface [51]. The lack of oxygen in mangrove creates an optimal environment for several groups of anaerobic organisms, such as sulfate-reducing bacteria [52], which could explain the presence of these groups in our samples. Although *Gammaproteobacteria*did not presented quantitative difference between depths, this class was one of the most abundant herein (36% of the total *Proteobacteria* sequences at both 10 cm and 40 cm depths). The dominance of *Gammaproteobacteria* in our samples corroborates the data reported by other studies that analyzed several different mangroves [8,49]. Some members of this phylum are important players of the ammonia oxidation [53]. Inside *Firmicutes*, the class *Clostridia* was dominant at 10 cm and *Bacilli* at 40 cm depth. Some members of this phylum are involved in the biodegradation of hydrocarbons [54], which are compounds commonly found in mangrove sediments, due to its proximity to urban areas and boat traffic [11,13].

Within the phylum *Verrucomicrobia*, the Subdivision 3 was dominant at 10 cm and the class *Verrucomicrobiae* was dominant at 40 cm depth. Members of this phylum are found in the rhizosphere of plants [55], and the class *Verrucomicrobiae* includes methanotrophic species [56]. In *Chloroflexi*, most of sequences from 10 cm belong to the class *Anaerolineae*, while at 40 cm depth 50% of sequences were unclassified at class level. Members of the phylum *Chloroflexi*play important role in the decomposition of organic matter [57]. Interestingly, the OM in our mangrove samples decreased in deeper layers and this phylum responded to its alteration by decreasing the abundance of sequences from 10 cm (2.8%) to 40 cm (0.75%) depth. Looking to the phylum *Bacteroidetes*, 52% of the sequences from 10 cm depth were not classified at any class level, while at 40 cm *Flavobacteria* was dominant. For the phylum *Acidobacteria*, for both depths the Group 18 was dominant, however at 10 cm the sequences were classified into 12 classes against only 3 in samples from 40 cm. From the total sequences obtained, approximately 33% were not classified into any bacterial phylum, which indicates that there is a vast diversity to be exploited in mangrove sediments.

The analysis of the bacterial variation in mangrove sediments indicates that the structure and composition of the communities are shaped by sediment characteristics, which change with site and depths. In our study, the depth of the sediment exerted a significant force in selecting different species inhabiting specific niches. Several studies have reported the relationship between sediment depth and microbial community structure [9,11,14,58,59].

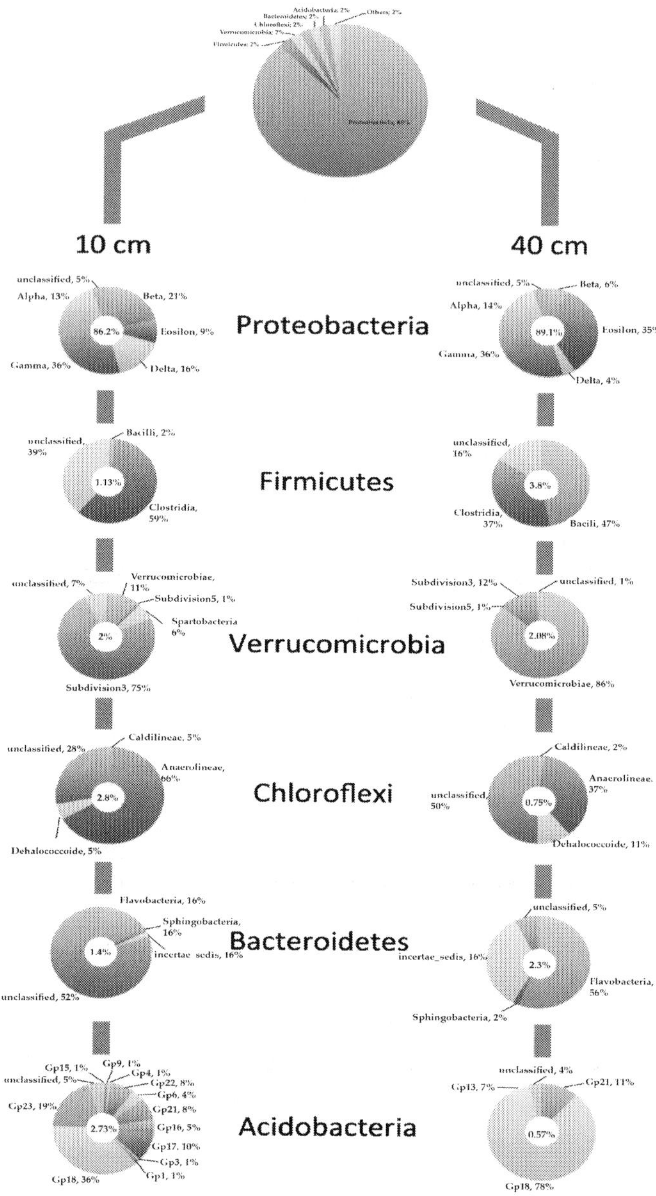

Figure 5: Bacterial composition distribution of partial 16S rRNA sequences from mangrove sediments at 10 and 40 cm depths. The sequences were classified using RDP Ribosomal Database Project classifier at phylum and class levels. The percentage inside circles refers to the relative abundance of each phylum at each depth.

CONCLUSIONS

In this study, the structure and composition of bacterial communities from mangrove sediments were described through fingerprinting and sequencing approaches. The bacterial communities displayed considerable diversity in shallow layers. Diversity decreased with increasing depth and its structure is significantly correlated with sediment characteristics. The bacterial composition shifted according to the depth and the affiliation of some groups suggests wide-scale participation in key nutrient cycles. In conclusion, our findings provide evidence that mangrove variations, such as nutrient content and depth, shape the microbial communities in the sediments and should be considered in studies of diversity. In recent decades, more interest has been given to the preservation of mangrove areas, due to its important functions performed in the environment. However, to be taken advantage of socio-economically, microbial activity should be known. Further studies on the microbial ecology in mangroves are required for a better understanding of the functional role of the microorganisms in these environments. This information will provide a better description of this complex system and the development of preservation strategies.

ACKNOWLEDGMENTS

This study was supported by a grant from Fundação de Amparo à Pesquisa do Estado de São Paulo (FAPESP 2007/01859-4 and FAPESP-Biota 2004/13910-6). The authors also thank Conselho Nacional de Desenvolvimento Científico e Tecnológico (CNPq) for a research fellowship to Siu Mui Tsai.

AUTHOR CONTRIBUTIONS

Lucas William Mendes and Siu Mui Tsai conceived and designed the experiment. Lucas William Mendes performed the experiments, analyzed the data and wrote the paper. Siu Mui Tsai was the higher supervisor and contributed reagents/materials/analysis tools.

REFERENCES

1. Nedwell, D.B.; Blackurn, T.H.; Wiebe, W.J. Dynamic nature of turnover of organic carbon, nitrogen and sulphur in the sediments of a Jamaican mangrove forest. *Mar. Ecol.: Prog. Ser.* 1994, *110*, 203–212.
2. Yan, B.; Hong, K.; Yu, Z.N. Archaeal communities in mangrove soil characterized by 16S rRNA gene clones. *J. Microbiol.* 2006, *44*, 566–571.
3. Lyimo, T.J.; Pol, A.; Jetten, M.S.M.; den Camp, H.J. Diversity of methanogenic Archaea in a mangrove sediment and isolation of a new *Methanococcoides* strain. *FEMS Microbiol. Lett.* 2009, *291*, 247–253.
4. Walsh, G.E. Mangroves: A review. In *Ecology of Halophytes*; Reinold, R.J., Queen, W.H., Eds.; Academic Press: New York, NY, USA, 1974; pp. 51–154.
5. Getter, C.D.; Cintron, G.; Dicks, B.; Lewis, R.R.; Seneca, E.D. The recovery and restoration of salt marshes and mangrove following an oil spill. In *Restoration of Habitats Impacted by Oil Spills*; Cairns, J., Buikema, A.L., Eds.; Butterworth: Boston, MA, USA, 1984; pp. 65–113.
6. Duke, N.C.; Meynecke, O.; Dittmann, S.; Ellison, A.M.; Anger, K.; Berger, U.; Cannicci, S.; Diele, K.; Ewel, K.C.; Field, C.D.; *et al.*,. A world without mangroves? *Sci. Lett.* 2007, *317*, 41–42.
7. Holguin, G.; Vazquez, P.; Bashan, Y. The role of sediment microorganisms in the productivity, conservation, and rehabilitation of mangrove ecosystems; an overview. *Biol. Fertil. Soils* 2001, *33*, 265–278.
8. Andreote, F.D.; Jiménez, D.J.; Chaves, D.; Dias, A.C.F.; Luvizotto, D.M.; Dini-Andreote, F.; Fasanella, C.C.; Lopez, M.V.; Baena, S.; Taketani, R.G.; *et al.*,. The microbiome of Brazilian mangrove sediments as revealed by metagenomics. *PLoS One* 2012, *7*, e38600.
9. Dias, A.C.F.; Andreote, F.D.; Rigonato, J.; Fiore, M.F.; Melo, I.S.; Araújo, W.L. The bacterial diversity in a Brazilian non-disturbed mangrove sediment. *Antonie Van Leeuwenhoek* 2010, *98*, 541–551.

10. Dias, A.C.F.; Dini-Andreote, F.; Taketani, R.G.; Tsai, S.M.; Azevedo, J.L.; Melo, I.S.; Andreote, F.D. Archaeal communities in the sediments of three contrasting mangroves. *J. Soils Sediments* 2011, *11*, 1466–1476.
11. Mendes, L.W.; Taketani, R.G.; Navarrete, A.A.; Tsai, S.M. Shifts in phylogenetic diversity of archaeal communities in mangrove sediments at different sites and depths in southeastern Brazil. *Res. Microbiol.* 2012, *163*, 366–377.
12. Sá, A.L.B.; Dias, A.C.F.; Qecine, M.C.; Cotta, S.R.; Fasanella, C.C.; Andreote, F.D.; Melo, I.S. Screening of endoglucanase-producing bacteria in the saline rhizosphere of Rhizophora mangle. *Braz. J. Microbiol.* 2014, *45*, 193–197.
13. Gomes, N.C.; Borges, L.R.; Paranhos, R.; Pinto, F.N.; Mendonca-Hagler, L.C.S.; Smalla, K. Exploring the diversity of bacterial communities in sediments of urban mangrove forests. *FEMS Microbiol. Ecol.* 2008, *66*, 96–109.
14. Taketani, R.G.; Franco, N.O.; Rosado, A.S.; van Elsas, J.D. Microbial community response to a simulated hydrocarbon spill in mangrove sediments. *J. Microbiol.* 2010, *48*, 7–15.
15. Sousa, O.V.; Macrae, A.; Menezes, F.G.; Gomes, N.C.; Vieira, R.H.; Mendonça-Hagler, L.C.S. The impact of shrimp farming effluent on bacterial communities in mangrove waters, Ceará, Brazil. *Mar. Pollut. Bull.* 2006, *52*, 1725–1734.
16. Margulies, M.; Egholm, M.; Altman, W.E.; Attiya, S.; Bader, J.S.; Bemben, L.A.; Berka, J.; Braverman, M.S.; Chen, Y.-J.; Chen, Z. *Nature* 2005, *437*, 376–380.
17. Van Raij, B.; Cantarella, H.; Andrade, J.C.; Quaggio, J.A. *Análise Química Para Avaliação da Fertilidade de Solos Tropicais*; Instituto Agronômico: Campinas, Brazil, 2001; p. 285.
18. Brody,J.R.;Kern,S.E.Sodiumboricacid:Atriz-less,coolerconductive medium for DNA electrophoresis. *Biotechniques* 2004, *36*, 214–216.
19. Amann, R.I.; Ludwig, W.; Schleifer, K.H. Phylogenetic identification and *in situ* detection of individual microbial cells without cultivation. *Microbiol. Rev.* 1995, *59*, 143–169.
20. Culman, S.W.; Bukowski, R.; Gauch, H.G.; Cadillo-Quiroz, H.;

Buckley, D.H. T-REX: Software for the processing and analysis of T-RFLP data. *BMC Bioinforma.* 2009, *10*, 171.

21. Culman, S.W.; Gauch, H.G.; Blackwood, C.B.; Thies, J.E. Analysis of T-RFLP data using analysis of variance and ordination methods: A comparative study. *J. Microbiol. Methods* 2008, *75*, 55–63.
22. Hammer, Ø.; Harper, D.A.T.; Ryan, P.D. PAST: Paleontological statistics software package for education and data analysis. *Palaeontol. Electron.* 2001, *4*, 9.
23. Sul, W.J.; Cole, J.R.; Jesus, E.D.C.; Wang, Q.; Farris, R.J.; Fish, J.A.; Tiedje, J.M. Bacterial community comparisons by taxonomy-supervised analysis independent of sequence alignment and clustering. *Proc. Natl. Acad. Sci. USA*2011, *108*, 14637–14642.
24. DeSantis, T.Z.; Hugenholtz, P.; Larsen, N.; Rojas, M.; Brodie, E.L.; Keller, K.; Huber, T.; Dalevi, D.; Hu, P.; Andersen, G.L. Greengenes, a chimera-checked 16S rRNA gene database and workbench compatible with ARB. *Appl. Environ. Microbiol.* 2006, *72*, 5069–5072.
25. Schloss, P.D.; Westcott, S.L.; Ryabin, T.; Hall, J.R.; Hartmann, M.; Hollister, E.B.; Lesniewski, R.A.; Oakley, B.B.; Parks, D.H.; Robinson, C.J.; *et al.*,. Introducing mothur: Open-source, platform-independent, community-supported software for describing and comparing microbial communities. *Appl. Environ. Microbiol.* 2009, *75*, 7537–7541.
26. Schloss, P.D.; Larget, B.R.; Handelsman, J. Integration of Microb Ecol statistics: A test to compare gene libraries. *Appl. Environ. Microbiol.* 2004, *70*, 5485–5492.
27. Varon-Lopez, M.; Dias, A.C.F.; Fasanella, C.C.; Durrer, A.; Melo, I.S.; Kuramae, E.E.; Andreote, F.D. Suplhur-oxidizing and sulfate-reducing communities in Brazilian mangrove sediments. *Environ. Microbiol.* 2014, *16*, 845–855.
28. Alongi, D.M. Bacterial productivity and microbial biomass in tropical mangrove sediments. *Microb. Ecol.* 1988, *15*, 59–79.
29. Holguin, G.; Zamorano, P.G.; Bashan, L.E.D.; Mendoza, R.; Amador, E.; Bashan, Y. Mangrove health in an arid envi-ronment encroached by urban development—A case study. *Sci. Total Environ.* 2006, *363*, 260–274.

30. Fierer, N.; Jackson, R.B. The diversity and biogeography of soil bacterial communities. *Proc. Natl. Acad. Sci. USA*2006, *103*, 626–631.

31. Jesus, E.D.; Marsh, T.L.; Tiedje, J.M.; Moreira, F.M.D. Changes in land use alter the structure of bacterial communities in Western Amazon soils. *ISME J.* 2009, *3*, 1004–1011.

32. Nielsen, U.N.; Osler, G.H.R.; Campbell, C.D.; Burslem, D.F.R.P.; van der Wal, R. The influence of vegetation type, soil properties and precipitation on the composition of soil mite and microbial communities at the landscape scale. *J. Biogeogr.* 2010, *37*, 1317–1328.

33. Hu, Y.J.; Xiang, D.; Veresoglou, S.D.; Chen, F.L.; Chen, Y.L.; Hao, Z.P.; Zhang, X.; Chen, B.D. Soil organic carbon and soil structure are driving microbial abundance and community composition across the arid and semi-arid grassland in northern China. *Soil Biol. Biochem.* 2014, *77*, 51–57.

34. Parmar, P.; Sindhu, S.S. Potassium solubilization by rhizosphere bacteria: Influence of nutritional and environmental conditions. *J. Microbiol. Res.* 2013, *3*, 25–31.

35. Richards, J.E.; Bates, T.E. Studies on the potassium-supplying capacities of southern Ontario soil. III. Measurement of available K. *Can. J. Soil Sci.* 1989, *69*, 597–610.

36. Basak, B.B.; Biswas, D.R. Influence of potassium solubilizing microorganism (*Bacillus mucilaginosus*) and waste mica on potassium uptake dynamics by sudan grass (*Sorghun vulgare* Pers.) grown under two Alfisols. *Plant Soil*2009, *317*, 235–255.

37. Belay, A.; Claassens, A.S.; Wehner, F.C. Effect of direct nitrogen and potassium and residual phosphorus fertilizers on soil chemical properties, microbial components and maize yield under long-term crop rotation. *Biol. Fertil. Soils* 2002, *35*, 420–427.

38. Pan, Y.; Cassman, N.; Hollander, M.; Mendes, L.W.; Korevaar, H.; Geerts, R.H.E.M.; van Veen, J.A.; Kuramae, E.E. Impact of long-term N, P, K, and NPK fertilization on the composition and potential functions of the bacterial community in grassland soil. *FEMS Microbiol. Ecol.* 2014, *90*, 195–205.

39. Fierer, N.; Schimel, J.P.; Holden, P.A. Variations in microbial

community composition through two soil depth profiles. *Soil Biol. Biochem.* 2003, *35*, 167–176.

40. Wilms, R.; Kopke, B.; Sass, H.; Chang, T.S.; Cypionka, H.; Engelen, B. Deep biosphere-related bacteria within the subsurface of tidal flat sediments. *Environ. Microbiol.* 2006, *8*, 709–719.
41. Pazinato, J.M.; Paulo, E.N.; Mendes, L.W.; Vazoller, R.F.; Tsai, S.M. Amazonian wetland soil and culture-dependent isolation of methanogenic Archaea. *Diversity* 2010, *2*, 1026–1047.
42. Li, N.; Yao, S.H.; You, M.Y.; Zhang, Y.L.; Qiao, Y.F.; Zou, W.X.; Han, X.Z.; Zhang, B. Contrasting development of soil microbial community structure under no-tilled perennial and tilled cropping during early pedogenesis of a Mollisol. *Soil Biol. Biochem.* 2014, *77*, 221–232.
43. Uroz, S.; Buée, M.; Murat, C.; Frey-klett, P.; Martin, F. Pyrosequencing reveals a contrasted bacterial diversity between oak rhizosphere and surrounding soil. *Environ. Microbiol. Rep.* 2010, *2*, 281–288.
44. Lemos, L.; Fulthorpe, R.; Triplett, E.W.; Roesch, L.F. Rethinking microbial diversity analysis in the high troughput sequencing era. *J. Microbiol. Methods* 2011, *86*, 42–51.
45. Ghosh, A.; Dey, N.; Bera, A.; Tiwari, A.; Sathyaniranjan, K.B.; Chakrabarti, K.; Chattopadhyay, D. Culture independent molecular analysis of bacterial communities in the mangrove sediment of Sudarban, India. *Saline Syst.* 2010, *6*, 1.
46. Gyaneshwar, P.; Hirsch, A.M.; Moulin, L.; Chen, W.M.; Elliott, G.N.; Bontemps, C.; Estrada-de los Santos, P.; Gross, E.; dos Reis, F.B.; Sprent, J.I.; *et al.,*. Legume-nodulating betaproteobacteria: Diversity, host range, and future prospects. *Mol. Plant–Microbe Interact.* 2011, *24*, 1276–1288.
47. Foti, M.; Sorokin, D.Y.; Lomans, B.; Mussman, M.; Zacharova, E.E.; Pimenov, N.V.; Kuenen, J.G.; Muyzer, G. Diversity, activity, and abundance of sulfate-reducing bacteria in saline and hypersaline Soda Lakes. *Appl. Environ. Microbiol.* 2007, *73*, 2093–2100.
48. Hori, T.; Muller, A.; Igarashi, Y.; Conrad, R.; Friedrich, M.W. Identification of iron-reducing microorganisms in anoxic rice paddy soil by ^{13}C-acetate probing. *ISME J.* 2010, *4*, 267–278.

49. Dos Santos, H.F.; Cury, J.C.; do Carmo, F.L.; dos Santos, A.L.; Tiedje, J.M.; van Elsas, J.D.; Rosado, A.S.; Peixoto, R.S. Mangrove bacterial diversity and the impact of oil contamination revealed by pyrosequencing: Bacterial proxies for oil pollution. *PLoS One* 2011, *6*, e16943.

50. Takai, K.; Campbell, B.J.; Cary, S.C.; Suzuki, M.; Oida, H.; Nunoura, T.; Hirayama, H.; Nakagawa, S.; Suzuki, Y.; Inagaki, F.; *et al.*,. Enzymatic and genetic characterization of carbon and energy metabolisms by deep-sea hydrothermal chemolithoautotrophic isolates of Epsilonproteobacteria. *Appl. Environ. Microbiol.* 2005, *71*, 7310–7320.

51. Lyimo, T.J.; Pol, A.; Op den Camp, H.J.M. Sulfate reduction and methanogenesis in sediments of Mtoni mangrove forest, Tanzania. *AMBIO* 2002, *31*, 614–616.

52. Dar, S.A.; Kleerebezem, R.; Stams, A.J.M.; Kuenen, J.G.; Muyzer, G. Competition and coexistence of sulfate-reducing bacteria, acetogens and methanogens in a lab-scale anaerobic bioreactor as affected by changing substrate to sulfate ratio. *Appl. Environ. Microbiol.* 2008, *78*, 1045–1055.

53. Kowalchuk, G.A.; Stephen, J.R. Ammonia-oxidizing bacteria: A model for molecular microbial ecology. *Annu. Rev. Microbiol.* 2001, *55*, 485–529.

54. Das, K.; Mukherjee, A.K. Crude petroleum-oil biodegradation efficiency of Bacillus subtilis and Pseudomonas aeroginosa strains isolated from a petroleum-oil contaminated soil from North-East India. *Bioresour. Technol.* 2007, *98*, 1339–1345.

55. Chow, M.L.; Radomski, C.C.; McDermott, J.M.; Davies, J.; Axelrood, P.E. Molecular characterization of bacterial diversity in Lodgepole pine (*Pinus contorta*) rhizosphere soils from British Columbia forest soils differing in disturbance and geographic source. *FEMS Microbiol. Ecol.* 2002, *42*, 347–357.

56. Dunfield, P.F.; Yuryev, A.; Senin, P.; Smirnova, A.V.; Stott, M.B.; Hou, S.; Ly, B.; Saw, J.H.; Zhou, Z.; Ren, Y.; *et al.*,. Methane oxidation by an extremely acidophilic bacterium of the phylum Verrucomicrobia. *Nature* 2007, *450*, 879–882.

57. Yamada, T.; Sekiguchi, Y.; Imachi, H.; Kamagata, Y.; Ohashi, A.;

Harada, H. Diversity localization, and physiological properties of filamentous microbes belonging to *Chloroflexi* sub phylum in mesophilic and thermophilic methanogenic sludge granules. *Appl. Environ. Microbiol.* 2005, *71*, 7493–7503.

58. Wilms, R.; Sass, H.; Kopke, B.; Cypionka, H.; Engelen, B. Methane and sulfate profiles within the subsurface of a tidal flat are reflected by the distribution of sulfate-reducing bacteria and methanogenic archaea. *FEMS Microbiol. Ecol.* 2007, *59*, 611–621.

59. Bagwel, C.E.; Formolo, M.; Ye, Q.; Yeger, C.M.; Lyons, T.W.; Zhang, C.L. Direct analysis of sulfate reducing bacterial communities in gas hydrate-impacted marine sediments by PCR-DGGE. *J. Basic Microbiol.* 2009, *49*, 87–92.

9

CHAPTER

CULTURE-DEPENDENT AND -INDEPENDENT METHODS CAPTURE DIFFERENT MICROBIAL COMMUNITY FRACTIONS IN HYDROCARBON-CONTAMINATED SOILS

Franck O. P. Stefani, Terrence H. Bell, Charlotte Marchand, Ivan E. de la Providencia, Abdel El Yassimi, Marc St-Arnaud, Mohamed Hijri

Department of Biological Sciences, Centre sur la biodiversité, Université de Montréal, Montréal, Québec, Canada

ABSTRACT

Bioremediation is a cost-effective and sustainable approach for treating polluted soils, but our ability to improve on current bioremediation strategies depends on our ability to isolate microorganisms from these soils. Although culturing is widely used in bioremediation research and applications, it is unknown whether the composition of cultured isolates closely mirrors the indigenous microbial community from contaminated soils. To assess this, we paired culture-independent (454-pyrosequencing of total soil DNA) with culture-dependent (isolation using seven different growth media) techniques to analyse the bacterial and fungal communities from hydrocarbon-contaminated soils. Although bacterial and fungal rarefaction curves were saturated for both methods, only 2.4% and 8.2% of the bacterial and fungal OTUs, respectively, were shared between datasets. Isolated taxa increased the total recovered species richness by only 2% for bacteria and 5% for fungi. Interestingly, none of the bacteria that we isolated were representative of the major bacterial OTUs recovered by 454-pyrosequencing. Isolation of fungi was moderately more effective at capturing the dominant OTUs observed by culture-independent analysis, as 3 of 31 cultured fungal strains ranked among the 20 most abundant fungal OTUs in the 454-pyrosequencing dataset. This study is one of the most comprehensive comparisons of microbial communities from hydrocarbon-contaminated soils using both isolation and high-throughput sequencing methods.

INTRODUCTION

Over the past few decades, human activities related to petroleum consumption have led to massive releases of both aliphatic and aromatic hydrocarbons, making these compounds some of the most ubiquitous environmental pollutants on Earth [1–3]. Among these are the polycyclic aromatic hydrocarbons (PAHs), which are of particular concern, given their persistence in the environment (especially in soil) and the potent carcinogenic, mutagenic, and teratogenic effects that these compounds have on living organisms [4–6]. Many isolated strains of bacteria and fungi can degrade at least some components of hydrocarbon contaminants in culture [7–11], which makes these two major soil microbial groups promising reservoirs of hydrocarbon-degrading activity.

A number of studies have shown that bioremediation, the use of living organisms to decontaminate polluted sites, is likely a feasible solution for treating these contaminants [11–13]. One approach to enhancing hydrocarbon bioremediation is the stimulation of indigenous hydrocarbon degraders, by supplying limiting nutrients, oxygen, and/or improving the physicochemical conditions of the polluted soil [13–15]. Alternatively, cultured hydrocarbon degraders can be used to degrade contaminants *ex situ*, or spiked into contaminated soils *in situ*, a process known as bioaugmentation [16–18]. Regardless of the approach taken, effective bioremediation hinges on our ability to study microbes that are indigenous to polluted sites. Culture-independent and-dependent methods for microbial community analysis have both been used frequently to describe microorganisms from hydrocarbon-contaminated environments.

For culture-independent analysis, high-throughput sequencing techniques [19,20] have revolutionized our view of microbial ecology, revealing hyperdiverse communities that are extremely responsive to hydrocarbon contaminants across a variety of environments [21–25]. Although culture-dependent methods generally recover a small portion of the diversity from soil environments, they are still a critical component of bioremediation development and research [26–30]. In addition to the potential *in situ* and *ex situ* applications of cultured isolates, microbial isolation allows *in vitro* assessments of isolate physiology and hydrocarbon degradation pathways and performance, providing a basis for annotating extensive metagenomic datasets, and helping to identify genes and/or organisms that could be useful in land reclamation.

Hydrocarbon-contaminated soils may be more amenable to comprehensive culture-dependent sampling than other soil environments, since hydrocarbon contamination often leads to a decline in microbial diversity [22,24,31,32], meaning that a lower sampling effort may be required to isolate a representative proportion of the active community. Hydrocarbon contaminants may also suppress certain sensitive groups [33] and tend to select primarily for subgroups of the *Actinobacteria* and *Proteobacteria* in affected soils [22,23]. Although these phyla are extremely diverse, they are some of the best represented among cultured isolates [34,35]. However, although the gap between culture-independent and culture-dependent analyses of soil microbial communities is often mentioned, few studies [36–38] have directly compared these approaches, and none has specifically investigated the biases associated with culturing both bacteria and fungi in a bioremediation context, despite the critical role of culturing in this field.

In this study, we used both nutrient-rich and impoverished media, supplemented with various types and concentrations of petroleum hydrocarbons, in order to assess the effectiveness of culture-based methods at recovering indigenous microorganisms from hydrocarbon-contaminated soils. We paired 454-pyrosequencing of bacterial 16S rDNA and the fungal ITS region with extensive culturing of bacteria and fungi, using sediment samples harvested from a basin that is highly contaminated with hydrocarbons, at the site of a former petrochemical plant. Since different substrates select for different groups of bacteria [39] and fungi [40], bacterial and fungal strains were isolated using seven different culture media to enhance the number of potential isolates. Although there are many approaches for cultivating soil bacteria and fungi [41], we chose to use basic nutrient-rich and impoverished agar plates, since these are still the most widely applied culturing method in microbiology. While we expected much lower richness in the cultured dataset, it was interesting to observe that current culturing methods do not capture most of the dominant microorganisms found in hydrocarbon-contaminated soils through 454-pyrosequencing. This was more apparent among the bacterial isolates than the fungal isolates. Even more surprisingly, a number of the cultured microorganisms were not identified at all in the 454-pyrosequencing data.

MATERIALS AND METHODS

Ethics Statement

No specific permits were required for the described field study. The land on which we conducted the phytoremediation field is privately owned by ConocoPhillips. ConocoPhillips gave permission for the study to be conducted on their land. This field study did not involve endangered or protected species.

Experimental Design and Sampling

Sampling occurred at the site of a former petrochemical plant at Varennes, on the south shore of the St-Lawrence River near Montreal, Quebec, Canada (45°41'56"N, 73°25'43"W). The sampling site is contaminated by a variety of industrial waste products related to petrochemical processing that have been released over the last forty years. The site was permanently closed in 2008, and since 2010, it has been used extensively to study the potential of willow cultivars for the phytoremediation of hydrocarbons (for details see www.genorem.ca and [23,42]). Five plots of 300 m^2 each were set up within a contaminated area of approximately 2500 m^2. Each plot was subdivided into 12 sub-plots of 25 m^2 each. In June 2011, two soil samples were collected at depths of 25 cm and 50 cm in each of the 12 sub-plots within each plot. The 24 soil samples from each plot were then pooled to obtain representative composite soil samples. Soil samples were chilled at 4°C during transport from the field to the lab, and were stored at -20°C until isolation and DNA extraction. A portion of each composite soil sample was analyzed for F1-F4 hydrocarbons (sum of all aromatic and aliphatic hydrocarbon compounds with chain lengths of C10-C50) by Maxxam Analytics (Montreal, Quebec, Canada) on June 2011, according to the protocol set forth by The Canadian Council of Ministers of the Environment. Results from hydrocarbon analyses (S1 Table) revealed an increasing contamination gradient from plots 1 through 5, which led us to classify the plots into three discrete contaminant levels: slightly contaminated (plots SC1, SC2), contaminated (plot C3), and highly contaminated (plots HC4, HC5).

Culture-Dependent (CD) Sample Processing

For each composite soil sample, one gram of soil was suspended in 9 ml of sterile distilled water and vortexed thoroughly. From this stock solution, serial dilutions were performed to 10^{-7}. Aliquots of 100 μl from dilutions of 10^{-3} and 10^{-4} for fungi, and of 10^{-6} and 10^{-7} for bacteria, were plated in duplicate on each of the seven culture media. Bacteria were isolated on tryptic soy agar (TSA, containing 30 g / L of tryptic soy broth (TSB)) and impoverished TSA plates (1 to 15 g /L of TSB) containing various concentrations of diesel engine oil or crude oil, or that had been coated with crude oil (Table 1). Fungal strains were isolated on potato dextrose agar (PDA, containing 24 g/L of potato dextrose broth (PDB)) and impoverished PDA plates (1 to 12 g/L of PDB) containing various concentrations of diesel engine oil or crude oil, or that had been coated with crude oil (Table 2). Petri dishes were inoculated, inverted, and incubated at 27°C for bacteria and 25°C for fungi. Bacterial colonies were checked every 48 hours for six days, and each new morphotype was subcultured on TSA for 48 hours. Fungal colonies were checked every 5 days for two weeks, and were then subcultured on PDA for three weeks. Bacterial and fungal subcultures were stored at 4°C until DNA isolation.

Table 1: Composition of the standard and selective media (based on one litre of medium) used to isolate soil bacteria

Each medium contained 100 mg / L of cycloheximide.

https://doi.org/10.1371/journal.pone.0128272.t001

Medium	Agar (g)	TSB (g)	Micro-element [d] (ml)	Macro-element [e] (ml)	Diesel (ml)	Oil (ml)	Crude Oil (ml)	Acetone (ml)
1.5% TSA—1%OD [a]	15	15			5	5		5
0.1% TSA—COC [b]	15	1	1	10			0.2	0.1
0.1% TSA—COM [c]	15	1	1	10			20	10
0.1% TSA—1%OD	15	1	1	10	5	5		5
0.1% TSA—5%OD	15	1	1	10	25	25		25
0.1% TSA—10%OD	15	1	1	10	50	50		50

[a] Oil–Diesel engine oil (Rotella diesel engine oil, Shell, Montreal, QC).

[b] Crude Oil was Coated onto the medium after solidification (from Gulf of Mexico, provided by Montreal pipeline).

[c] Crude Oil was mixed with the growth medium (Montreal pipeline).

[d] $MgSO_4$ (739 mg / L), KNO_3 (76 mg / L), KCl (65 mg / L), KH_2PO_4 (4.1 mg / L).

[e] $MnSO_4*7H_2O$ (6 mg / L), $ZnSO_4*7H_2O$ (2.65 mg / L), H_3BO_3 (1.5 mg / L). $CuSO_4$ (0.13 mg / L), $Na_2MoO_4*2H_2O$ (0.002 mg / L), KI (0.75 mg / L).

doi:10.1371/journal.pone.0128272.t001

Table 2: Composition of the standard and selective media (based on one liter of medium) used to isolate soil fungi

https://doi.org/10.1371/journal.pone.0128272.t002

Medium	Agar (g)	PDB (g)	Micro-element (ml)	Macro-element (ml)	Diesel (ml)	Oil (ml)	Crude Oil (ml)	Acetone (ml)
1.2% PDA—1%OD	15	12			5	5		5
0.1% PDA—COC	15	1	1	10			0.2	0.1
0.1% PDA—COM	15	1	1	10			20	10
0.1% PDA—1%OD	15	1	1	10	5	5		5
0.1% PDA—5%OD	15	1	1	10	25	25		25
0.1% PDA—10%OD	15	1	1	10	50	50		50

Each medium contained 200 mg / L of streptomycin, and 100 mg / L of ampicillin.

doi:10.1371/journal.pone.0128272.t002

DNA Isolation, Amplification, and Sequencing of Microbial Isolates

Bacterial strains were picked with a 1 µl sterile inoculation loop (Sarstedt, Montreal, Canada) and spiked directly into a PCR master mix (described below). Fungal strains were subcultured for one week in PDB before we harvested fresh mycelium for the isolation of genomic DNA (gDNA). All gDNA isolations were performed using a Freedom EVO100 extraction robot (Tecan Group, Mannedorf, Switzerland) with the NucleoMag 96 Plant kit (Macherey Nagel, Oesingen, Switzerland) according to the manufacturer's instructions. Bacterial 16S rDNA and fungal ITS sequences were amplified using the 27F [43] / 1492R [44], and the ITS1-F [45] / ITS4 [46] primer sets, respectively. PCR mixtures were made up of 1× PCR buffer, 4% DMSO, 0.5 mg BSA, 0.5 mM $MgCl_2$, 0.2 mM of each deoxynucleotide triphosphate, 0.5 μM of each primer, one unit of DreamTaq DNA polymerase (Fermentas, Canada) and one μl of gDNA (or picked colony) in a total volume of 50 μl. Thermal cycling conditions for bacteria were as follows: initial denaturation at 94°C for 5 min; 30 cycles at 94°C for 1 min, 55°C for 1 min, and 72°C for 1 min; and a final elongation at 72°C for 10 min. Thermal cycling conditions for fungi were as follows: initial denaturation at 94°C for 2 min; 35 cycles at 94°C for 30 s, 52°C for 30 s, and 72°C for 1 min; and a final elongation at 72°C for 5 min. PCR reactions were performed using an Eppendorf Mastercycler ProS (Eppendorf, Mississauga, ON), and products were visualized on GelRed-stained 1.5% agarose gels using the Gel-Doc system (Bio-Rad Laboratories, Mississauga, ON). DNA sequencing was performed on an Applied Biosystems 3730xl DNA analyzer (Applied Biosystems, Carlsbad, CA) at the McGill University and Genome Quebec Innovation Centre (Montreal, QC). Sanger sequencing data are deposited in GenBank under the accession numbers KP177318—KP177405 and KP177406—KP177454 for bacteria and fungi, respectively.

Culture-Independent (CI) Sample Processing

Total soil DNA was isolated from one gram of each composite soil sample using the MoBio PowerSoil DNA Isolation Kit following the manufacturer's instructions. For each sample, triplicate PCR reactions of partial 16S rDNA and fungal ITS amplicons were performed using barcoded primers with the required 454 adapter sequences (S2A and S2B Table). Thermal cycling conditions for reactions using 16S rDNA barcoded primers were as follows: initial denaturation at 95°C for 5 min; 30 cycles at 95°C for 30 s, 55°C for 30 s, and 72°C for 30 s; and a final elongation at 72°C for 7 min. Thermal cycling conditions for reactions using ITS barcoded primers were as follows: initial denaturation at 94°C for 4 min; 30 cycles at 94°C for 30 s, 50°C for 60 s, and 72°C for 90 s; and a final elongation at 72°C for 10 min. PCR reactions were performed on an Eppendorf Mastercycler ProS (Eppendorf, Mississauga, ON). PCR triplicates were combined and quantified using the Quant-iT PicoGreen dsDNA assay kit (Invitrogen, Life Technologies). 16S rDNA and ITS pools from each of the five composite samples were then combined in equimolar ratios. High-throughput sequencing was performed on the 454

GS FLX+ platform using Lib-L chemistry (Roche, Branford, CT, USA). The 16S rDNA and ITS pools were each sequenced using ¼ of a plate. The 454-pyrosequencing data generated in this study have been deposited in the NCBI Sequence Read Archive and are available under the BioSample accession numbers SAMN03199986 and SAMN03199987.

Bioinformatic Analyses

Quality processing (read quality trimming, chimera check) of partial 16S rDNA and ITS1 sequences was performed in Mothur v.1.33.2 following the protocol used in Schloss et al. [47] and accessed online (http://www.mothur.org/wiki/454_SOP) in September 2014. Sequences obtained from Sanger sequencing were edited, cleaned, and assembled in Geneious Pro v.6.1.5 (Biomatters). For both 454 sequences and Sanger sequences, clustering analyses were performed in Geneious, and OTUs were defined at 97% similarity. For 454-pyrosequencing datasets, singletons were not considered, and doubletons with a nucleotide similarity <100% were excluded. OTU rarefaction curves, Venn diagrams, taxonomic identifications, relative abundance analyses, and richness and diversity indices were performed in Mothur. Taxonomic identifications for bacterial and fungal sequences were performed in Mothur using the SILVA [48] and UNITE [49] databases, respectively. Prior to performing community comparisons based on the 454-pyrosequencing datasets, each library was randomly subsampled to the minimum number of sequences observed per composite soil sample. Heatmaps and Venn diagrams were created using the R statistical language v.3.0.0 [50] with the packages gplots [51] and VennDiagram [52]. Krona charts were calculated using the KronaTools available from http://krona.sourceforge.net [53]. To partition the datasets between abundant and rare OTUs, the number of reads observed for each OTU was divided by the number of reads counted for the most abundant OTU. Rare OTUs were defined as those with a read proportion of < 5% (S1 Fig).

RESULTS

After quality trimming, OTU filtering, and standardizing the number of sequences per sample, 22,422 16S rDNA sequences and 10,098

ITS sequences were recovered from bacterial and fungal CI datasets, respectively (Table 3). OTU richness reached saturation at each level of contamination (S2A and S2B Fig), with the exception of plot 3 in the fungal CD dataset (coverage = 59.2%, Table 3). Using a similarity threshold of 97% to cluster sequences within the same OTU, a total of 2047 and 360 OTUs were recovered from bacterial and fungal CI datasets, respectively. For the culture-dependent (CD) datasets, 781 bacterial strains and 279 fungal strains were isolated and sequenced from the five plots, using the seven different culture media per microbial group. A total of 88 bacterial and 49 fungal OTUs were defined based on analysis of the 16S rDNA and ITS sequences, respectively.

Table 3: Bacterial and fungal OTU richness and diversity recovered in each library

https://doi.org/10.1371/journal.pone.0128272.t003

Method [a]	Level of contamination [b]	No. sequences	Coverage [c] (%)	Nb. OTUs	D [d]	H [e]	PD [f]
CI-Bacteria	SC	9082	98.7	1342	658.4	6.8	57.9
	C	4519	94.1	1030	385.9	6.4	47.3
	HC	8821	98.6	887	135.1	5.9	47.2
CI-Fungi	SC	4017	99.5	235	35.1	4.3	32.1
	C	2033	98.5	144	16.5	3.6	19.4
	HC	4048	99.3	153	17.1	3.5	26.5
CD-Bacteria	SC	297	91.2	56	16.8	3.3	2.46
	C	172	93.6	30	9.0	2.7	1.36
	HC	312	98.7	36	13.0	3.0	1.47
CD-Fungi	SC	223	94.2	43	19.8	3.2	5.42
	C	27	59.2	16	15.2	2.5	2.35
	HC	29	89.6	8	4.3	1.6	1.12

[a] CI: culture-independent (454-pyrosequencing); CD: culture-dependent (isolation).

[b] SC: slightly contaminated; C: contaminated; HC: highly contaminated (see S1 Table).

[c] Good's coverage.

[d] Simpson's inverse index of diversity.

[e] Shannon index.

[f] Faith's phylogenetic diversity.

doi:10.1371/journal.pone.0128272.t003

Bacterial and Fungal Communities recovered by CI and CD Methods

The CI and CD methods provided two extremely divergent views of the microbial communities recorded in the five plots (Fig 1). The proportion of shared OTUs between CD and CI datasets was only 2.4% (51 OTUs) for bacteria and 8.2% (31 OTUs) for fungi (Fig 1A and 1B). This did, however, represent 58% and 63% of the bacterial and fungal strains isolated *in vitro*. Based on Bray-Curtis dissimilarity, bacterial and fungal communities from each contaminant level clustered first according to sampling method (due to the large difference in species richness between CI and CD datasets), and secondarily according to the level of hydrocarbon-contamination (Fig 1C and 1D), with the single exception of the CI HC fungal community which segregated independently from the other fungal assemblages. The heatmaps clearly show that the most abundant bacterial and fungal OTUs recovered via culturing do not correspond with those recovered by 454-pyrosequencing. *Proteobacteria* was the dominant bacterial phylum, representing 61% of the 16S rDNA reads from the 197 most abundant bacterial OTUs (Fig 2A). *Rhodocyclales* (15%), *Burkholderiales* (10%), *Actinomycetales* (10%), *Rhizobiales* (8%), *Xanthomonadales* (7%), and *Sphingomonadales* (5%) were the most frequently observed bacterial orders in the CI sequences (Fig 2A). Among the 51 bacterial OTUs shared between CI and CD methods (Fig 2B), OTU7 (*Arthrobacter* sp.) was the most frequently isolated, but represented the 61st most abundant OTU out of 2047 in the CI dataset. Most of the shared OTUs belong to the genera *Sphingobium* and *Sphingomonas* (*Sphingomonodales*), *Pseudomonas* (*Pseudomonadales*) and *Arthrobacter* (*Actinomycetales*). *Sphingomonodales*and *Actinomycetales* were the most abundant orders identified within the shared OTUs, with relative abundances of 33% and 30% respectively, while accounting for 5% and 10% of reads in the CI dataset.

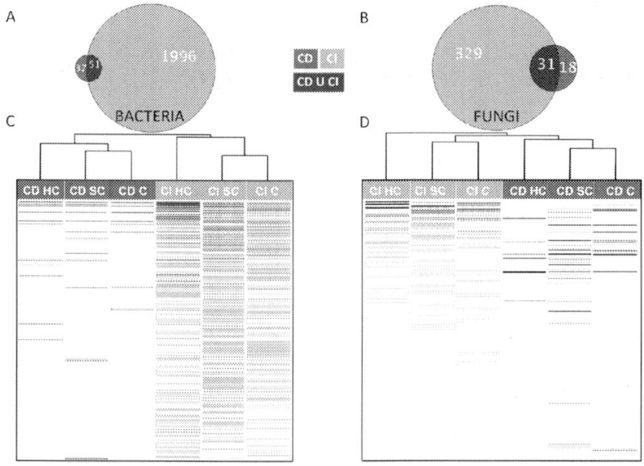

Figure 1: Proportional Venn diagram showing the distribution of bacterial (A) and fungal (B) OTUs between culture-independent (CI, light grey) and culture-dependent (CD, dark grey) methods.

Heatmap distribution of the relative abundance of bacterial (C) and fungal (D) OTUs recorded with CI (light grey) and CD (dark grey) methods. OTUs are in rows and colour intensity indicates relative abundance, with black indicating the highest relative abundances observed for slightly contaminated (SC), contaminated (C) and highly contaminated (HC) composite soil sediments recovered with CI and CD methods. OTUs are presented by descending number of reads from top to bottom, based on proportion of total reads recorded in CI and CD datasets. The top dendrogram shows the hierarchical clustering of bacterial and fungal communities recovered for each PAH contamination level based on Bray-Curtis dissimilarity.

https://doi.org/10.1371/journal.pone.0128272.g001

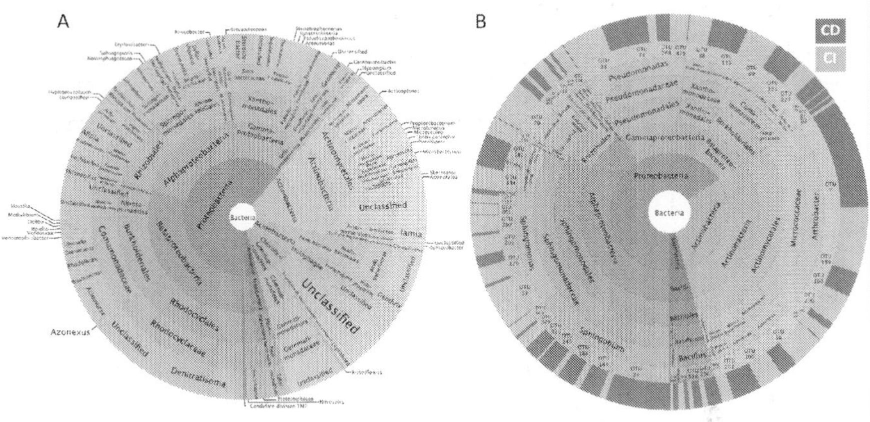

Figure 2: Krona charts showing the taxonomic identification and relative abundance of the most abundant bacterial OTUs recorded in CI datasets (A) and shared OTUs between CI and CD datasets (B).

The proportion of sequences from the CI (light grey) or CD (dark grey) datasets for each shared OTU is shown in the outer ring.

https://doi.org/10.1371/journal.pone.0128272.g002

With respect to the fungal communities observed in the CI dataset, *Ascomycota* was the dominant phylum, representing 65% of all reads (Fig 3A), while no *Glomeromycota* sequence was recovered in the CI dataset. At the order level, *Agaricales* (13%), *Saccharomycetales*(12%), *Hypocreales* (10%), *Pleosporales* (9%), *Sordariales* (8%) and

Spizellomycetales (6%) dominated the fungal community. Eighteen percent of the reads could not be identified. In contrast to the bacterial datasets, some fungal isolates ranked among the 20 most abundant OTUs in the CI dataset (Fig 3B). For example, OTU7, OTU13, and OTU15 were ranked seventh, thirteenth, and sixteenth in the CI dataset, respectively, based on read abundance. Only two isolated strains (OTU 75 and 87) were more frequently observed in the CD dataset than in the CI dataset (Fig 3B). A large proportion of the shared OTUs could not be identified (39% of all shared sequences), while *Hypocreales* (40%) and *Pleosporales* (10%) were the two prevalent fungal orders in the CD dataset. While some members of the *Basidiomycota* were recorded in the CI dataset, none were isolated.

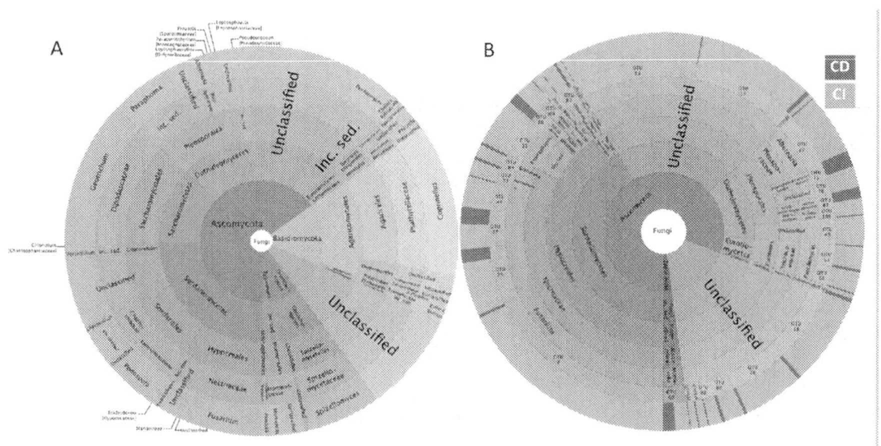

Figure 3: Krona charts showing the taxonomic identification and relative abundance of the most abundant fungal OTUs recorded in CI datasets (A) and shared OTUs between CI and CD datasets (B).

The proportion of sequences from the CI (light grey) or CD (dark grey) datasets for each shared OTU is shown in the outer ring.

https://doi.org/10.1371/journal.pone.0128272.g003

Interestingly, some of the isolated bacterial and fungal strains could not be detected by 454-pyrosequencing (S3A and S3B Fig), despite the fact that the number of reads from the CI analysis was more than an order of magnitude higher than the number of sequences in CD datasets, and the rarefaction curves were saturated. *Sphingobium* represented 24% of the bacterial taxa that were found only in the CD dataset, followed by *Staphylococcus* (14%) and *Arthrobacter* (8%) (S3A Fig). For fungi, OTU81 and OTU108 (*Trichoderma*), and OTU97 and OTU118 (*Paraphoma*) represented 66% of the isolated strains that were not detected by 454-pyrosequencing (S3B Fig). The majority of non-shared bacterial and fungal OTUs were isolated on media containing oil-derived hydrocarbons (data not shown).

DISCUSSION

Culturing is essential to both the study and application of microbially-mediated bioremediation, yet the proportion and identity of the microbial community in hydrocarbon-contaminated soils that is culturable has remained unclear. Our results show that not only does culturing select a small proportion of the microorganisms that can be observed through CI methods, a fact that has been long known, but that many of the most abundant microorganisms detected using CI methods are not captured with plate-culturing techniques. This suggests that the microbes that are selected by culturing are not even representative of those that are highly competitive in the environment.

Culture-Independent versus Culture-Dependent Methods

As shown by saturated rarefaction curves, the sampling effort for both CI and CD methods was adequate for capturing most of the abundant OTUs that could be identified with these methods in the five composite sediment samples. This allows us to draw robust comparisons between these datasets. Previous studies that have compared bacterial or fungal diversity using CI and CD approaches concluded that these methods were complementary [36,38,54–57]. In this study, the number of taxa that were unique to the CD dataset was limited, despite the use of seven different culture media for isolation. Taxa identified from isolation alone increased total species richness by only 2% for bacteria and 5% for fungi, meaning that 454-pyrosequencing captured 95% or more of the microbial diversity that could be obtained using these methods, and with far less effort. The OTUs that were only identified by culturing likely represent rare taxa that could not be observed at the sequencing depth used in this study. Extremely rare taxa may in fact represent a large proportion of the OTUs in natural microbial communities [58]. The presence of hydrocarbons in six out of the seven isolation media could explain the limited number of OTUs that we recovered through isolation, since hydrocarbon contaminants can negatively affect soil bacterial and fungal diversity [22,23,59]. Alternative culturing methods, such as the use of rhizosphere isolation medium [37], micro-cultivation techniques [40,60] and dilute nutrient media [25,61,62], have isolated a larger number of microorganisms, but the overlap with CI community analysis has been limited in all cases, and many OTUs that are highly abundant in CI datasets cannot be isolated. For instance Shade et al. [37] performed 454-pyrosequencing on total cultivated soil bacteria without colony picking, and reported that only 9.1% of bacterial OTUs were shared between CD and CI datasets, while each dataset contained 37,645 and 46,347 reads, representing a total of 4957 OTUs. As in our study, the authors demonstrated that many of the shared OTUs represented rare taxa in the CI dataset. Lagier et al. [36] investigated human gut bacteria by microbial culturomics (212 different culture conditions, yielding 32,500 colonies) and metagenomics, and in this case, only 5.1% of total OTUs were shared between sampling approaches. Furthermore, the authors identified 174 novel bacterial species in the CD dataset, demonstrating that metagenomics can be an inefficient method for detecting extremely rare bacteria.

Since many of the bacterial and fungal taxa that we isolated are not representative of the most abundant taxa *in situ*, their potential as effective *in situ* inoculants is questionable, even if some of these isolates demonstrated a high capacity for hydrocarbon degradation (unpublished data). Intermicrobial competition appears to play an important role in shaping microbial activity and abundance [63–65], and the persistence and activity of bioaugmented strains depends fundamentally on their ability to compete with indigenous microorganisms [66]. If strains isolated through culturing are only transient contributors to microbial community function, or exist permanently within the rare biosphere, they are unlikely to dominate when reinoculated, even if their abundance is maximized in the inoculum. This may be a major contributing factor to the poor efficiency of many prior attempts at bioaugmentation [66,67]. Spatial and temporal monitoring of microbial abundance in contaminated soils using CI methods appears to be a fundamental step towards determining the factors that govern the abundance of microorganisms *in situ*, as well as for selecting strains that can be successful across varied environments. However, if members of the rare biosphere are more efficient at hydrocarbon degradation than microbes that are naturally abundant, it may be desirable to develop vacant niches that allow these organisms to persist. This may be possible through the addition of plant hosts, for example [68].

Comparing Fungal and Bacterial Isolation from Hydrocarbon-Contaminated Soils

A greater proportion of fungal OTUs identified via CI analysis were isolated from the hydrocarbon-contaminated sediments when compared with isolation of bacteria. This is likely due in part to the lower richness of fungi than bacteria in soil. While roughly 1000 fungal OTUs were identified in 4 grams of forest soil [69], between 6000–50,000 bacterial species can potentially be recovered from a single gram of soil, depending on where the "species" cutoff is drawn [70]. In addition, fungi and bacteria may be differentially sensitive to small changes in environmental conditions. Fungal propagules, such as hyphae and spores, are generally much larger than bacteria. As a result, certain soil microenvironments that are relevant to bacterial distribution may not affect fungi, as these organisms may develop niches at different spatial scales. Fungi have also been shown to be much less sensitive to pH variation than bacteria, which might allow cultivation of a wider range of fungi than bacteria using media at only one pH level [71]. On the other hand, fungi have been shown to be more sensitive to high hydrocarbon concentrations and were more strongly influenced by the presence of plants than bacteria in this study [23].

Still, only a fraction of known fungal species can be cultured (ranging from 5 to 17% according to [72–74]), while most species that belong to the phylum *Basidiomycota* cannot yet be isolated or grown under *in vitro* conditions. Taxa within the *Glomeromycota* cannot be cultivated at all without host plant roots. In our CI dataset, two taxa from the *Agaricomycetes* (*Basidiomycota*) ranked first and second in terms of read abundance, but were not captured by culturing efforts. These two OTUs have non-conspecific matches with ITS sequences of *Coprinellus*, a genus in the family *Psathyrellaceae*. Their absence in the CD dataset is somewhat surprising, since species of *Coprinellus* and *Coprinopsis* have been grown in Petri dishes [75]. Suhara et al. [75] reported near-complete degradation of 1 μmol of polychlorinated dibenzo-p-dioxin by *Coprinellus disseminatus* in two weeks. Minimal information is available with regards to the hydrocarbon-degrading capacity of members of this family, although *Coprinellus bisporus* was recorded, along with other fungi, from soil contaminated with phenanthrene and pyrene and treated with pea wheat straw [76]. The third most abundant OTU had a non-conspecific match with ITS sequences from the genus *Geotrichum*. *Geotrichum candidum* was previously found to increase benzo[a]pyrene degradation in mineral liquid medium [77]. Zheng & Obbard [78] isolated a strain identified as *Geotrichum* sp. in a PAH-contaminated soil, which was able to oxidize more than 50% of the phenanthrene present in liquid medium after seven days. The authors were able to isolate and cultivate this strain of *Geotrichum* sp. using generic media that were not hydrocarbon-coated, while the OTU observed at our site was not isolated, despite the use of both selective and non-selective media.

The three most abundant bacterial taxa in our CI dataset had non-conspecific matches in GenBank with environmental 16S rDNA sequences from heavy metal or oil-contaminated soils. The first two OTUs belong to the *Rhodocyclaceae* (*Denitratisoma* and unclassified taxon) while the third is a member of the *Comamonadaceae* (*Acidovorax*). PAH-degrading bacteria have been reported from both of these families [79–81]. In addition, members of the *Rhodocyclaceae*have been shown to degrade PAHs adsorbed in hydrophobic sorbents, highlighting the ability of these strains to degrade PAHs even when they are poorly bioavailable [80]. Thus, the most abundant bacterial and fungal taxa recorded by 454-pyrosequencing from this hydrocarbon-contaminated site belong to taxonomic groups that include members with the capacity to degrade hydrocarbons. Unfortunately, none of these taxa could be isolated using traditional culturing techniques, preventing comprehensive testing of the hydrocarbon-degrading potential of these strains.

CONCLUSIONS

To date, this is the most comprehensive comparison of CI and CD methods in hydrocarbon-contaminated soils, and the first to examine fungal communities. Although microbial diversity can be reduced in contaminated environments, we did not recover a large proportion of the bacterial or fungal OTUs that were identified through 454-pyrosequencing. Surprisingly, many of the most abundant OTUs *in situ* were not cultured, despite the use of several types of hydrocarbon-containing media, suggesting that factors other than hydrocarbon tolerance and metabolism are critical to *in situ* community composition. Although obligate hydrocarbonoclastic microorganisms have been identified [82], they appear to be uncommon, and many microorganisms can adapt their existing metabolic pathways to the degradation of hydrocarbon compounds. Based on this, as well as our results, hydrocarbons may be a poor selective agent for comprehensively culturing representative microorganisms from hydrocarbon-contaminated soils, despite the widespread use of hydrocarbon-coated media for isolating strains from such environments. This suggests that novel culturing techniques are essential to the progression of bioremediation research. Although next-generation sequencing has caused CI analyses to wildly outpace our ability to culture microorganisms, novel 'omics' tools will be invaluable in developing future generations of culturing methods, as some of the environmental and ecological factors that shape microbial abundance can be identified and exploited.

ACKNOWLEDGMENTS

We thank Sébastien Halary and Yves Terrat for providing assistance with metagenomic data treatment. This work was supported by the Genome Quebec and Genome Canada funded GenoRem Project.

AUTHOR CONTRIBUTIONS

Conceived and designed the experiments: AEY MH MSA. Performed the experiments: CM AEY IEDLP. Analyzed the data: FS. Contributed reagents/materials/analysis tools: MH. Wrote the paper: FS THB IEDLP MSA MH.

REFERENCES

1. Echeveste P, Dachs J, Berrojalbiz N, Agustí S (2010) Decrease in the abundance and viability of oceanic phytoplankton due to trace levels of complex mixtures of organic pollutants. Chemosphere 81: 161–168. pmid:20673958
2. Samanta SK, Singh OV, Jain RK (2002) Polycyclic aromatic hydrocarbons: environmental pollution and bioremediation. TRENDS in Biotechnology 20: 243–248. pmid:12007492
3. Eckle P, Burgherr P, Michaux E (2012) Risk of large oil spills: a statistical analysis in the aftermath of deepwater horizon. Environ Sci Technol 46: 13002–13008. pmid:23127138
4. Billiard SM, Meyer JN, Wassenberg DM, Hodson PV, Di Giulio RT (2008) Nonadditive effects of PAHs on early vertebrate development: mechanisms and implications for risk assessment. Toxicol Sci 105: 5–23. pmid:18156145
5. Phillips DH (1983) Fifty years of benzo(a)pyrene. Nature 303: 468–472. pmid:6304528
6. Bispo A, Jourdain MJ, Jauzein M (1999) Toxicity and genotoxicity of industrial soils polluted by polycyclic aromatic hydrocarbons (PAHs). Organic Geochemistry 30: 947–952.
7. Kanaly RA, Harayama S (2000) Biodegradation of high-molecular-weight polycyclic aromatic hydrocarbons by bacteria. J Bacteriol 182: 2059–2067. pmid:10735846
8. Toledo FL, Calvo C, Rodelas B, González-López J (2006) Selection and identification of bacteria isolated from waste crude oil with polycyclic aromatic hydrocarbons removal capacities. Syst Appl Microbiol 29: 244–252. pmid:16564960
9. Das K, Mukherjee AK (2007) Crude petroleum-oil biodegradation efficiency of *Bacillus subtilis* and *Pseudomonas aeruginosa* strains isolated from a petroleum-oil contaminated soil from North-East India. Bioresource Technology 98: 1339–1345. pmid:16828284

10. Harms H, Schlosser D, Wick LY (2011) Untapped potential: exploiting fungi in bioremediation of hazardous chemicals. Nature Reviews Microbiology 9: 177–192. pmid:21297669
11. Aislabie J (2006) Bioremediation of hydrocarbon-contaminated polar soils. Extremophiles 10: 171–179. pmid:16514512
12. Vasudevan N, Rajaram P (2001) Bioremediation of oil sludge-contaminated soil. Environment International 26: 409–411. pmid:11392759
13. Liu W, Luo Y, Teng Y, Li Z, Ma LQ (2009) Bioremediation of oily sludge-contaminated soil by stimulating indigenous microbes. Environ Geochem Health 32: 23–29. pmid:19363671
14. Atlas RM (1995) Bioremediation of petroleum pollutants. International Biodeterioration & Biodegradation 35: 317–327. pmid:23637980
15. Ferrari BC, Zhang C, van Dorst J (2011) Recovering greater fungal diversity from pristine and diesel fuel contaminated sub-antarctic soil through cultivation using both a high and a low nutrient media approach. Front Microbiol 2: 217. pmid:22131985
16. Fan M-Y, Xie R-J, Qin G (2014) Bioremediation of petroleum-contaminated soil by a combined system of biostimulation–bioaugmentation with yeast. Environmental Technology 35: 391–399. pmid:24600879
17. Mishra S, Jyot J, Kuhad RC, Lal B (2001) Evaluation of inoculum addition to stimulate in situ bioremediation of oily-sludge-contaminated soil. Appl Environ Microb 67: 1675–1681. pmid:11282620
18. Salami S, Olusola A, Elum E, Anslem E (2010) Bioremediation of a crude oil polluted soil with *Pleurotus pulmonarius* and *Glomus mosseae* using *Amaranthus hybridus* as a test plant. J Bioremed Biodegrad 01.
19. Tringe SG, Hugenholtz P (2008) A renaissance for the pioneering 16S rRNA gene. Current Opinion in Microbiology 11: 442–446. pmid:18817891
20. Shokralla S, Spall JL, Gibson JF, Hajibabaei M (2012) Next-generation sequencing technologies for environmental DNA research. Mol Ecol 21: 1794–1805. pmid:22486820

21. Mason OU, Hazen TC, Borglin S, Chain PSG, Dubinsky EA, Fortney JL, et al. (2012) Metagenome, metatranscriptome and single-cell sequencing reveal microbial response to Deepwater Horizon oil spill. The ISME Journal 6: 1715–1727. pmid:22717885
22. Bell TH, Yergeau E, Maynard C, Juck D, Whyte LG, Greer CW (2013) Predictable bacterial composition and hydrocarbon degradation in Arctic soils following diesel and nutrient disturbance. The ISME Journal 7: 1200–1210. pmid:23389106
23. Bell TH, El-Din Hassan S, Lauron-Moreau A, Al-Otaibi F, Hijri M, Yergeau E, et al. (2014) Linkage between bacterial and fungal rhizosphere communities in hydrocarbon-contaminated soils is related to plant phylogeny. The ISME Journal 8: 331–343. pmid:23985744
24. Sutton NB, Maphosa F, Morillo JA, Abu Al-Soud W, Langenhoff AAM, Grotenhuis T, et al. (2012) Impact of long-term diesel contamination on soil microbial community structure. Appl Environ Microb 79: 619–630.
25. Vartoukian SR, Palmer RM, Wade WG (2010) Strategies for culture of "unculturable" bacteria. FEMS Microbiol Lett 309: 1–7. pmid:20487025
26. Wong J, Lai KM, Wan CK, Ma KK (2002) Isolation and optimization of PAH-degradative bacteria from contaminated soil for PAHs bioremediation. Water, Air, and Soil Pollution.
27. Watanabe K (2001) Microorganisms relevant to bioremediation. Current Opinion in Biotechnology 12: 237–241. pmid:11404100
28. Chaerun SK, Tazaki K, Asada R, Kogure K (2004) Bioremediation of coastal areas 5 years after the Nakhodka oil spill in the Sea of Japan: isolation and characterization of hydrocarbon-degrading bacteria. Environment International 30: 911–922. pmid:15196839
29. Potin O, Rafin C, Veignie E (2004) Bioremediation of an aged polycyclic aromatic hydrocarbons (PAHs)-contaminated soil by filamentous fungi isolated from the soil. International Biodeterioration & Biodegradation 54: 45–52. pmid:23637980
30. Malaviya P, Rathore V (2007) Bioremediation of pulp and paper mill effluent by a novel fungal consortium isolated from polluted soil. Bioresource Technology 98: 3647–3651. pmid:17208440

31. Saul DJ, Aislabie JM, Brown CE, Harris L, Foght JM (2005) Hydrocarbon contamination changes the bacterial diversity of soil from around Scott Base, Antarctica. FEMS Microbiol Ecol 53: 141–155. pmid:16329936
32. Bell TH, Joly S, Pitre FE, Yergeau E (2014) Increasing phytoremediation efficiency and reliability using novel omics approaches. TRENDS in Biotechnology 32: 271–280. pmid:24735678
33. Santos dos HF, Cury JC, do Carmo FL, Santos dos AL, Tiedje J, van Elsas JD, et al. (2011) Mangrove bacterial diversity and the impact of oil contamination revealed by pyrosequencing: bacterial proxies for oil pollution. PLoS ONE 6: e16943. pmid:21399677
34. Woese CR (1987) Bacterial evolution. Microbiological Reviews 51: 221–271. pmid:2439888
35. Hugenholtz P, Goebel BM, Pace NR (1998) Impact of culture-independent studies on the emerging phylogenetic view of bacterial diversity. J Bacteriol 180: 4765–4774. pmid:9733676
36. Lagier J-C, Armougom F, Million M, Hugon P, Pagnier I, Robert C, et al. (2012) Microbial culturomics: paradigm shift in the human gut microbiome study. Clin Microbiol Infect 18: 1185–1193. pmid:23033984
37. Shade A, Hogan CS, Klimowicz AK, Linske M, McManus PS, Handelsman J (2012) Culturing captures members of the soil rare biosphere. Environ Microbiol 14: 2247–2252. pmid:22788977
38. Pitkäranta M, Meklin T, Hyvärinen A, Paulin L, Auvinen P, Nevalainen A, et al. (2008) Analysis of fungal flora in indoor dust by ribosomal DNA sequence analysis, quantitative PCR, and culture. Appl Environ Microb 74: 233–244. pmid:17981947
39. Köpke B, Wilms R, Engelen B, Cypionka H, Sass H (2005) Microbial diversity in coastal subsurface sediments: a cultivation approach using various electron acceptors and substrate gradients. Appl Environ Microb 71: 7819–7830. pmid:16332756
40. Collado J, Platas G, Paulus B, Bills GF (2007) High-throughput culturing of fungi from plant litter by a dilution-to-extinction technique. FEMS Microbiol Ecol 60: 521–533. pmid:17466028
41. Pham VHT, Kim J (2012) Cultivation of unculturable soil bacteria.

TRENDS in Biotechnology 30: 475–484. pmid:22770837

42. Guidi W, Kadri H, Labrecque M (2012) Establishment techniques to using willow for phytoremediation on a former oil refinery in southern Quebec: achievements and constraints. Chemistry and Ecology 28: 49–64.
43. Lane DJ (1991) 16S/23S rRNA sequencing. In: Stackebrandt E, Goodfellow M, editors. Nucleic acid techniques in bacterial systematics (Modern microbiological methods). New York, NY: John Wiley and Sons. pp. 115–175.
44. Turner S, Pryer KM, Miao VPW (1999) Investigating deep phylogenetic relationships among cyanobacteria and plastids by small subunit rRNA sequence analysis. Journal of Eukaryotic Microbiology 46: 327–338. pmid:10461381
45. Gardes M, Bruns TD (1993) ITS primers with enhanced specificity for basidiomycetes—application to the identification of mycorrhizae and rusts. 2: 113–118. pmid:8180733
46. White T, Bruns T, Lee S, Taylor J (1990) Amplification and direct sequencing of fungal ribosomal RNA genes for phylogenetics. In Innis M A, Gelfand D H, Sninsky J J, and White T J (ed), PCR protocols: a guide to methods and applications Academic Press Inc, New York, NY: 315–322.
47. Schloss PD, Gevers D, Westcott SL (2011) Reducing the effects of PCR amplification and sequencing artifacts on 16S rRNA-based studies. PLoS ONE 6: e27310. pmid:22194782
48. Pruesse E, Quast C, Knittel K, Fuchs BM, Ludwig W, Peplies J, et al. (2007) SILVA: a comprehensive online resource for quality checked and aligned ribosomal RNA sequence data compatible with ARB. Nucleic Acids Res 35: 7188–7196. pmid:17947321
49. Abarenkov K, Henrik Nilsson R, Larsson K-H, Alexander IJ, Eberhardt U, Erland S, et al. (2010) The UNITE database for molecular identification of fungi–recent updates and future perspectives. New Phytol 186: 281–285. pmid:20409185
50. R Development Core Team (n. d.) R: A language and environment for statistical computing. Available: http://www. R-project. org/.
51. Warnes GR (2013) Includes R source code and/or documentation contributed by: Ben Bolker, Lodewijk Bonebakker, Robert

Gentleman, Wolfgang Huber Andy Liaw, Thomas Lumley, Martin Maechler, Arni Magnusson, Steffen Moeller, Marc Schwartz and Bill Venables (2013). gplots: Various R programming tools for plotting data. R package version 2. 11. 0. 1. http://CRAN. R-project. org/package=gplots.

52. Chen H, Boutros PC (2011) VennDiagram: a package for the generation of highly-customizable Venn and Euler diagrams in R. BMC Bioinformatics 12.
53. Ondov BD, Bergman NH, Phillippy AM (2011) Interactive metagenomic visualization in a Web browser. BMC Bioinformatics 12: 385. pmid:21961884
54. Barreto MC, Houbraken J, Samson RA, Brito D, Gadanho M, San Romão MV (2012) Unveiling the fungal mycobiota present throughout the cork stopper manufacturing process. FEMS Microbiol Ecol 82: 202–214. pmid:22630140
55. Carraro L, Maifreni M, Bartolomeoli I, Martino ME, Novelli E, Frigo F, et al. (2011) Comparison of culture-dependent and-independent methods for bacterial community monitoring during Montasio cheese manufacturing. Res Microbiol 162: 231–239. pmid:21262359
56. Vaz-Moreira I, Egas C, Nunes OC, Manaia CM (2011) Culture-dependent and culture-independent diversity surveys target different bacteria: a case study in a freshwater sample. Antonie Van Leeuwenhoek International Journal of General and Molecular Microbiology 100: 245–257.
57. Daghino S, Murat C, Sizzano E, Girlanda M, Perotto S (2012) Fungal diversity is not determined by mineral and chemical differences in serpentine substrates. PLoS ONE 7: e44233. pmid:23028507
58. Haegeman B, Hamelin J, Moriarty J, Neal P, Dushoff J, Weitz JS (2013) Robust estimation of microbial diversity in theory and in practice. The ISME Journal 7: 1092–1101. pmid:23407313
59. Pérez-Leblic MI, Turmero A, Hernández M, Hernández AJ, Pastor J, Ball AS, et al. (2012) Influence of xenobiotic contaminants on landfill soil microbial activity and diversity. J Environ Manage 95 Suppl: S285–S290. pmid:20724060
60. Shrestha P, Szaro TM, Bruns TD, Taylor JW (2011) Systematic

search for cultivatable fungi that best deconstruct cell walls of *Miscanthus* and sugarcane in the field. Appl Environ Microb 77: 5490–5504. pmid:21685162

61. Stewart EJ (2012) Growing unculturable bacteria. J Bacteriol 194: 4151–4160. pmid:22661685
62. Janssen PH, Yates PS, Grinton BE, Taylor PM, Sait M (2002) Improved culturability of soil bacteria and isolation in pure culture of novel members of the divisions *Acidobacteria*, *Actinobacteria*, *Proteobacteria*, and *Verrucomicrobia*. Appl Environ Microb 68: 2391–2396. pmid:11976113
63. Cornforth DM, Foster KR (2013) Competition sensing: the social side of bacterial stress responses. Nature Reviews Microbiology 11: 285–293. pmid:23456045
64. Foster KR, Bell T (2012) Competition, not cooperation, dominates interactions among culturable microbial species. Current Biology 22: 1845–1850. pmid:22959348
65. Hibbing ME, Fuqua C, Parsek MR, Peterson SB (2009) Bacterial competition: surviving and thriving in the microbial jungle. Nature Reviews Microbiology 8: 15–25.
66. Thompson IP, van der Gast CJ, Ciric L, Singer AC (2005) Bioaugmentation for bioremediation: the challenge of strain selection. Environ Microbiol 7: 909–915. pmid:15946288
67. Tyagi M, da Fonseca MMR, de Carvalho CCCR (2011) Bioaugmentation and biostimulation strategies to improve the effectiveness of bioremediation processes. Biodegradation 22: 231–241. pmid:20680666
68. Gentry TJ, Rensing C, Pepper IL (2004) New approaches for bioaugmentation as a remediation technology. Critical Reviews in Environmental Science and Technology 34: 447–494.
69. Buée M, Reich M, Murat C, Morin E, Nilsson RH, Uroz S, et al. (2009) 454 Pyrosequencing analyses of forest soils reveal an unexpectedly high fungal diversity. New Phytol 184: 449–456. pmid:19703112
70. Curtis TP, Sloan WT, Scannell JW (2002) Estimating prokaryotic diversity and its limits. P Natl Acad Sci Usa 99: 10494–10499. pmid:12097644

71. Rousk J, Bååth E, Brookes PC, Lauber CL, Lozupone C, Caporaso JG, Knight R, Fierer N. (2010) Soil bacterial and fungal communities across a pH gradient in an arable soil. The ISME Journal 4: 1340–1351. pmid:20445636

72. Bridge P, Spooner B (2001) Soil fungi: diversity and detection. Plant Soil 232: 147–154.

73. O'Brien H, Parrent J, Jackson J, Moncalvo J, Vilgalys R (2005) Fungal community analysis by large-scale sequencing of environmental samples. Appl Environ Microb 71: 5544–5550. pmid:16151147

74. Hawksworth D (2001) The magnitude of fungal diversity: the 1. 5 million species estimate revisited. Mycol Res 105: 1422–1432.

75. Suhara H, Kamei I, Maekawa N, Kondo R (2011) Biotransformation of polychlorinated dibenzo- p-dioxin by *Coprinellus* species. Mycoscience 52: 48–52.

76. Shahsavari E, Adetutu EM, Anderson PA, Ball AS (2013) Necrophytoremediation of phenanthrene and pyrene in contaminated soil. J Environ Manage 122: 105–112. pmid:23567029

77. Rafin C, de Foucault B, Veignie E (2013) Exploring micromycetes biodiversity for screening benzo[a]pyrene degrading potential. Environ Sci Pollut Res 20: 3280–3289.

78. Zheng Z, Obbard JP (2003) Oxidation of polycyclic aromatic hydrocarbons by fungal isolates from an oil contaminated refinery soil. Environ Sci Pollut Res 10: 173–176. pmid:12846378

79. Martin F, Torelli S, Le Paslier D, Barbance A, Martin-Laurent F, Bru D, et al. (2012) *Betaproteobacteria* dominance and diversity shifts in the bacterial community of a PAH-contaminated soil exposed to phenanthrene. Environmental Pollution 162: 345–353. pmid:22243884

80. Regonne RK, Martin F, Mbawala A, Ngassoum MB, Jouanneau Y (2013) Identification of soil bacteria able to degrade phenanthrene bound to a hydrophobic sorbent in situ. Environ Pollut 180: 145–151. pmid:23770314

81. Mbadinga SM, Wang L-Y, Zhou L, Liu J-F, Gu J-D, Mu B-Z (2011) Microbial communities involved in anaerobic degradation of alkanes. International Biodeterioration & Biodegradation 65: 1–13. pmid:23637980

82. Yakimov MM, Timmis KN, Golyshin PN (2007) Obligate oil-degrading marine bacteria. Current Opinion in Biotechnology 18: 257–266. pmid:17493798

10

CHAPTER

THE POTENTIAL ROLE OF ARBUSCULAR MYCORRHIZAL FUNGI IN THE RESTORATION OF DEGRADED LANDS

Fisseha Asmelash[1,2], Tamrat Bekele[2], and Emiru Birhane[3,4]

[1] Forest and Range Land Biodiversity Conservation Directorate, Ethiopian Biodiversity Institute, Addis Ababa, Ethiopia

[2] Department of Plant Biology and Biodiversity Management, Addis Ababa University, Addis Ababa, Ethiopia

[3] Department of Land Resources Management and Environmental Protection, Mekelle University, Mekelle, Ethiopia

[4] Department of Ecology and Natural Resource Management, Norwegian University of Life Sciences, Ås, Norway

ABSTRACT

Experiences worldwide reveal that degraded lands restoration projects achieve little success or fail. Hence, understanding the underlying causes and accordingly, devising appropriate restoration mechanisms is crucial. In doing so, the ever-increasing aspiration and global commitments in degraded lands restoration could be realized. Here we explain that arbuscular mycorrhizal fungi (AMF) biotechnology is a potential mechanism to significantly improve the restoration success of degraded lands. There are abundant scientific evidences to demonstrate that AMF significantly improve soil attributes, increase above and belowground biodiversity, significantly improve tree/shrub seedlings survival, growth and establishment on moisture and nutrient stressed soils. AMF have also been shown to drive plant succession and may prevent invasion by alien species. The very few conditions where infective AMF are low in abundance and diversity is when the soil erodes, is disturbed and is devoid of vegetation cover. These are all common features of degraded lands. Meanwhile, degraded lands harbor low levels of infective AMF abundance and diversity. Therefore, the successful restoration of infective AMF can potentially improve the restoration success of degraded lands. Better AMF inoculation effects result when inocula are composed of native fungi instead of exotics, early seral instead of late seral fungi, and are consortia instead of few or single species. Future research efforts should focus on AMF effect on plant community primary productivity and plant competition. Further investigation focusing on forest ecosystems, and carried out at the field condition is highly recommended. Devising cheap and ethically widely accepted inocula production methods and better ways of AMF *in situ* management for effective restoration of degraded lands will also remain to be important research areas.

Keywords: AMF, ecological restoration, facilitation, inoculation, land degradation, mycorrhiza, monoxenic culture, succession

INTRODUCTION

Ecological restoration has emerged to be the central theme of global environmental policies (Aradottir and Hagen, 2013; Jacobs et al., 2015). Restoration of at least 15% of the world's degraded ecosystems is one of the 20 2011–2020 targets of the UN Convention on Biological Diversity

(CBD, 2010[1]). In 2011, world leaders endorsed the "Bonn challenge" which is a global commitment to restore 150 million hectares of deforested and degraded lands by 2020 (Aradottir and Hagen, 2013). In 2014, the New York Declaration on Forests put forward even a bigger global commitment of restoring 350 million hectares of deforested and degraded lands until 2030 (Jacobs et al., 2015). Most importantly, in 2015, the UN concretized these global commitments by adopting the 2030 Sustainable Development Goals which has one of the 17 targets (Target 15) dealing on ecological restoration (UN, 2015). However, restoration experiences so far show that many restoration projects achieve limited success or fail completely (Thomas et al., 2014) and therefore, extra effort is needed to achieve the huge global restoration commitments put on the table. Here we propose AMF inoculation and *in situ* management for better restoration outcome of degraded lands.

About 93% of flowering plant families (Brundrett, 2009) and 92% of land plant families (Wang and Qiu, 2006) are estimated to have mycorrhizal associations. These associations, based on their structure and physiological relationship with symbionts, are categorized in to seven; of which arbuscular mycorrhiza is one (Brundrett et al., 1996; Barea et al., 2011). Arbuscular mycorrhiza is the most predominant and evolutionarily the ancestor of all the association types (Wang and Qiu, 2006). AMF produce arbuscules, hyphae, and vesicles within host plants' root cortical cells (Brundrett and Abbott, 2002). However, some species within the family Gigasporaceae do not form vesicles but instead, form auxiliary cells of unknown function (Redecker and Raab, 2006). In few other cases as well, arbuscules develop poorly or may be absent (Koide and Mosse, 2004). AMF are; (1) obligate biotrophs completely depending on host plants for organic carbon (File et al., 2012), (2) evolutionarily intimately associated with plants (Taylor et al., 1995), (3) multiple nucleated, and (4) asexually reproducing eukaryotes (Schüßler et al., 2007).

Arbuscular mycorrhizal fungi are keystone organisms with myriads of ecosystem roles. The external hyphae network (extraradical mycelium) of the fungi permeate in to the microsites of rocks and soils surrounding the plant roots (Finlay, 2008; Barea et al., 2011) increasing the root absorbing surface area 100 or even 1000 fold (Larcher, 1995). Therefore, AMF increase plants' nutrient and water relation (e.g., Birhane et al., 2012, 2015; Banerjee et al., 2013), and can improve plants' field survival and establishment (e.g., Pouyu-Rojas and Siqueira, 2000; Habte et al., 2001; Ouahmane et al., 2006; Dag et al., 2009; Kapulnik et al., 2010; Karthikeyan and Krishnakumar, 2012; Manaut et al., 2015). AMF improve soil structure, soil water relation, plants' tolerance to biotic and abiotic stresses, increase plants' nutrient supply, plants' growth, yield and reproductive success and reduce fertilizer requirement (Finlay, 2008; Gianinazzi et al., 2010; Simard and Austin, 2010; Barea et al., 2011; Al-Karaki, 2013; Soka and Ritchie, 2014). AMF influence plant community structure (Van der Heijden et al., 1998; Hartnett and Wilson, 1999; Renker et al., 2004; Heneghan et al., 2008; Lin et al., 2015) and are considered to have a pivotal role in plant community assembly and succession (Janos, 1980; Renker et al., 2004; Kikvidze et al., 2010). Therefore, AMF have significant role in ecological restoration.

The potential role of AMF in ecological restoration has been well recognized even before restoration ecology emerged as a scientific field of study (see Janos, 1980 and the references there). However, as of yet, there is no report available to confirm that AMF inoculation has grown to be a biotechnological tool that is widely applicable in ecological restoration. Review articles dealing on the subject are also very few. To our knowledge, those review articles that dealt on the subject are Skujins and Allen (1986), Brundrett and Abbott (2002), Jeffries et al.,. (2002) and Renker et al.,. (2004). Other reviews (e.g., Perry et al., 1987; Quoreshi, 2008; Sanon et al., 2010; Al-Karaki, 2013) did not deal on AMF specifically and the one by Koide and Mosse (2004) allocated some paragraphs for the topic. Therefore, although there is a large number of articles and sufficient knowledge on the ecosystem role of AMF, their role in the restoration of degraded lands is relatively little reviewed. Hence, the purpose of this review article is to gather data from published articles and assess the effects AMF have on measurable ecological restoration attributes and ecological processes.

FEATURES OF DEGRADED LANDS

There is no single internationally approved definition of land degradation (World Resource Institute, 2015[2]). However, land degradation is often defined as a long-term loss of ecosystem function and productivity caused by disturbances from which the land cannot recover unaided (e.g., Bai et al., 2008). Meanwhile, reduction in net primary productivity has commonly been used to measure the level of land degradation and restoration (Bai et al., 2008). The Society for Ecological Restoration (SER), however, recommends nine attributes to measure restoration success (SER, 2004). Earlier, Aronson et al.,. (1993) adapted Odum (1969) succession traits to formulate their own restoration attributes. The SER restoration attributes are excellent parameters (Ruiz-Jaen and Aide, 2005), however, it is the Aronson et al.,. (1993) restoration attributes that have commonly been used by restoration ecologists (Choi, 2004; Ruiz-Jaen and Aide, 2005). Therefore, in this article, the attributes listed by Aronson et al.,. (1993) are adapted and used to safely characterize degraded lands (Table 1).

Table 1: Features of degraded lands compared to reference climax ecosystems (Based on Aronson et al., 1993)

Structural indicators of degraded lands	Functional indicators of degraded lands	
Low total plant cover	Low biomass productivity	
Low perennial and annual plant species richness	Low soil organic matter	
Low aboveground phytomass	Poor soil water relation	Lowered soil water reserves
Low beta diversity (species turnover along environmental gradient)		Low coefficient of rainfall efficiency (the amount of water infiltrating to middle and deep soil layers)
Decreased life form spectrum (Decreased number of species with different modes of adaptation)		Reduced length of water availability period
Reduced number of keystone species	Low rain use efficiency (RUE)	
Low soil microbial biomass	Poor nutrient cycling index (the ratio of the amount of nutrients mainly N&P recycled to the amount leaching or lost)	
Low soil microbial diversity	Low nitrogen use efficiency (NUE)	

Degraded lands are characterized by low levels of AMF abundance and diversity. An experiment carried out in Brazil (Cardozo-Junior et al., 2012) compared the abundance and diversity of AMF on lands of differing degradation levels and also determined the same on a young restoring site. The result of the observation clearly showed that as the scale of degradation increases, the abundance and diversity of AMF reduces and when restoration presumes both AMF abundance and diversity increase (Cardozo-Junior et al., 2012). Elsewhere also, it was reported that an effective exclosure increased AMF abundance (Birhane et al., 2010). These are in agreement to the remarks by Abbott and Robson (1991) and Schnoor et al.,. (2011) which indicated that although AMF are ubiquitous, the very few conditions where a natural ecosystem can be devoid of AMF is in areas that are severely eroded or disturbed.

A greenhouse experiment with simulated erosion was able to demonstrate that erosion of soil beyond 7.5 cm could make soil loose AMF completely (Habte, 1989). Likewise, Jasper et al.,. (1989) were able to experimentally determine that, while AMF maintained their infective potential in extremely dry soil conditions, their infective potential was significantly lowered when the soil was disturbed. Hyphae are important source of inoculum but are highly susceptible to disturbance and hence, disturbance leads to lowered infective potential of AMF (Brundrett and Abbott, 2002).

Furthermore, several studies conducted in agricultural fields have shown that disturbance not only reduces AMF abundance, diversity and infectivity but also results in drastic shift in the AMF community (Schnoor et al., 2011). Most species of the most common AMF families (Glomeraceae, Acaulosporaceae, and Gigasporaceae) have distinctive biomass allocation strategies whereby species of the Glomeraceae allocate most of their biomass in the intraradical hyphae while species of the Gigasporaceae allocate most of their biomass in the extraradical hyphae and species of the Acaulosporaceae produce low biomass both intra and extraradically (Maherali and Klironomos, 2007). Similarly, these distinctive fungal groups have distinctive life history with most species of the Glomeraceae being ruderals while that of Gigasporaceae and Acaulosporaceae are competitors and stress tolerators respectively (Chagnon et al., 2013). Ruderal AMF species are disturbance tolerant since they have shorter extraradiacal mycelium and have the following life history strategy viz., grow faster, have short life cycle and invest earlier and more abundantly in spore formation, fuse fragmented hyphae more readily, and form cross-walls that enable infected root pieces and severed hyphal fragments to heal and re-colonize host roots (Chagnon et al., 2013). Meanwhile, AMF communities of disturbed sites are characteristically dominated by disturbance tolerant species of the family Glomeraceae and more specifically the genus Glomus (Chagnon et al., 2013).

The number of surviving propagules of AMF in soils also declines with time in the absence of host plants (Brundrett and Abbott, 2002). Alexander et al.,. (1992) have reported that heavy logging in a Malaysian forest significantly reduced (75% reductions) the abundance and infectivity of AMF propagules. Therefore, considering the fact that land degradation significantly reduces plant cover, increases soil disturbance and erosion, low levels of AMF abundance, diversity and infective potential can be considered as a peculiar feature of degraded lands. Degraded lands are also prone to invasion by exotic alien species. This is because, the low level of native plants diversity can potentially provide vacant niche for invasives (Mack et al., 2000).

WHAT IS ECOLOGICAL RESTORATION?

Ecological restoration, according to Hobbs et al.,. (2007) is a process of assembly and succession mediated by disturbance. Succession refers to the more or less regular and predictable replacement of seral communities while community assembly refers to the species dynamics of each seral community. Community assembly is typically viewed as a hierarchical process with local species assemblages representing subsets of a larger species pool (Kikvidze et al., 2015). There are three level filters that result in a particular seral community assemblages viz. (1) speciation, extinction and migration, (2) dispersal, and (3) habitat filters (abiotic factors) and biotic filters like competition and facilitation (Gotzenberger et al., 2011). A typical community assembly was observed by Gleason (1927) whereby ponds at similar locality with similar environmental condition formed different kinds of wetland communities.

Restoration strategies of degraded terrestrial systems usually center on manipulation of species order of arrival and modification of filters to accelerate succession and/or jump start succession (Young et al., 2001; Hobbs et al., 2007; Gómez-Aparicio, 2009). The seral community assemblage is very important in ecological succession since it can determine the latter seral community assemblage and hence, succession trajectory. Egler (1954), argued that ecosystem development can be accelerated by controlling initial species composition and succession to achieve the desired end point. Some field observations (e.g., Cortines and Valcarcel, 2009) have shown that this has practical application in ecological restoration. This phenomenon is known as the priority effect (Young et al., 2005; Zedler, 2005). Hence, designing initial species composition and ensuring their survival and establishment is an important step in ecological restoration (Angelini et al., 2011). With the growing appreciation to plant–plant facilitation interaction and due to better adaptation to resource limited conditions, early/mid successional shrubs are gaining preference to late successional tree/shrubs to startup restoration process of degraded lands (Gómez-Aparicio, 2009; Padilla et al., 2009).

Restoration ecologists not only provide appropriate conditions for desired species to establish but they also, in the meantime, devise ways of preventing the establishment by invasives (Zedler, 2005). Disturbance is also an important factor in ecological restoration since it can modify filters and community assemblage (Choi, 2004). Restoration ecologists have observed that some low level of natural disturbance (e.g., logging, fire, flooding, etc.) can enhance biological diversity and hence, ecological restoration (Palmer et al., 1997).

Based on the scope and complexity of intervention, ecological restoration ranges from species reintroduction to population restoration to community restoration (Young et al., 2001). Based on the restoration goal, it ranges from reclamation to rehabilitation to true restoration. An ecosystem that is slightly disturbed can restore back to its pre-disturbance status (true restoration). As the magnitude of disturbance increases the return to pre-disturbance status may be impossible and hence, return to an intermediate successional status of the given community (an alternative steady state) may be achieved (rehabilitation). When the disturbance is severe, the threshold of irreversibility is passed and the return to pre-disturbance community status or intermediate successional status will be completely impossible and hence, restoration can only result in a novel community stature (reclamation) (Aronson et al., 1993). In the advent of climate change, to have reclamation as a restoration goal is considered to be relevant since novel climatic conditions are anticipated in the future (Choi, 2004).

Considering the wide range of concepts embedded in ecological restoration, as shown above, a comprehensive definition is crucial. Hence, the SER defined ecological restoration as, the process of assisting the recovery of an ecosystem that has been degraded, damaged, or destroyed (SER, 2004). This is the most widely accepted definition of ecological restoration (Harris et al., 2006; Higgs et al., 2014). Likewise, for this article, this definition is adopted.

In tropical lands ecological restoration, tree planting (Lamb et al.,

2005; Holl et al., 2010; Aerts and Honnay, 2011) and re-vegetation/ reforestation (Cortines and Valcarcel, 2009; Al-Karaki, 2013) are known to be the most effective and widely used biological measures. Accordingly, in this article ecological restoration is considered to be the re-vegetation of degraded sites mainly through tree/shrub planting.

AMF AND THE MYCORRHIZOSPHERE ECOLOGY

The rhizosphere is a narrow zone of soil affected by the presence of plant roots (Hrynkiewicz and Baum, 2011). It is extremely important and active area for root activity and metabolism (Saharan and Nehra, 2011). Roots release a multitude of organic compounds (e.g., exudates and mucilage) derived from photosynthesis and other plant processes making the rhizosphere a hot spot of microbial activities mainly that of fungi and bacteria (Hrynkiewicz and Baum, 2011). The physical, chemical and biological environment of the rhizosphere is hence, clearly distinct from the bulk soil (Barea et al., 2002).

Similarly, the rhizosphere of the mycorrhizal plant can be referred to as the mycorrhizosphere (Barea et al., 2002). Mycorrhizosphere comprises both the root and hyphae influence zones or the rhizosphere and hyphosphere (Timonen and Marschner, 2006). Mycorrhizal hyphal growth in soils is extensive, with mycelial lengths reaching 111 m cm^{-3} or 0.5 mg g^{-1} or 900 kg ha^{-1} of soil (Simard and Austin, 2010). Hence, the mycorrhizosphere provide a critical link between plants, other microorganisms and the soil (Hrynkiewicz and Baum, 2011).

Intricate interactions take place within the mycorrhizosphere. The most important ones could be interactions between; AMF and the plant, AMF and bacteria, AMF and other fungi, and among AMF themselves. These interactions commence when plant roots exude strigolactones (SLs) (Parniske, 2008; Gutjahr, 2014). Under phosphate or nitrogen limiting conditions plants exude elevated amounts of SLs into the rhizosphere (Gutjahr, 2014). SLs are carotenoid-derived plant hormones (Gutjahr, 2014) that induce AMF spore germination and hyphal branching (Parniske, 2008). They are also known to induce seed germination in parasitic plants, such as *Striga* (Parniske, 2008) and are also involved in suppression of shoot branching and shaping root architecture (Gutjahr, 2014). The AMF on their part, produce mycorrhiza (Myc) factors. Myc-factors induce calcium oscillations in root epidermal cells and also activate plant symbiosis-related genes (Parniske, 2008). Then the AMF form special type of appressoria called hyphopodia which develops from mature hyphae (Parniske, 2008). As a consequence of sequential chemical and mechanical stimulation, plant epidermal cells produce a pre-penetration apparatus (PPA) (Parniske, 2008). Subsequently, a fungal hypha that extends from the hyphopodium enters the PPA, which guides the fungus through root cells toward the cortex. The fungus then leaves the plant cell and enters the apoplast, where it branches and grows laterally along the root axis (Parniske, 2008; Gutjahr, 2014). These hyphae induce the development of PPA-like structures in inner cortical cells, subsequently enter these cells, and branch to form arbuscules (Parniske, 2008). Upon getting nourished via the arbuscules the fungi will develop extraradical mycelium whose leading tips form new spores to continue the lifecycle of the fungi (Parniske, 2008). Vesicles, which are proposed to function as storage organs of the fungus, when applicable, are formed in the apoplast (Parniske, 2008).

Arbuscular mycorrhizal fungi have a significant role in plants' P nutrition and sometimes, 100% of the P may be provided by the AMF (Smith et al., 2011). In return, plants allocate, according to most authors, 4–20% of the photosynthate to the AMF (Lerat et al., 2003). It has been shown that plants preferentially allocate more carbon in favor of the more beneficial fungi (e.g., Bever et al., 2009). Moreover, plants allow the arbuscules to live in their cells as long as the AMF is delivering phosphorous and maybe other nutrients efficiently (Parniske, 2008). The observation that mutation of the arbuscule specific phosphate transporter PT4 results in premature degradation of arbuscules suggests that the lifetime of arbuscules is influenced by their ability to deliver phosphate and probably other nutrients (Parniske, 2008). This provides the plant with a means to maintain efficient arbuscules and penalize inefficient ones with early degradation. Conceptually, this mechanism allows the plant not only to discriminate between efficient and inefficient fungal species but also allows to remove potentially 'good' fungal symbionts that are attached to a poor phosphate source. This concept allows fungal clones and species to compete for arbuscule formation, which allows succession in an established root system (Parniske, 2008). Meanwhile, the formation of fungal colonization structures and the extent of root colonization are largely under plant control (Gutjahr, 2014).

Arbuscular mycorrhizal fungi are known to play role in plant nutrition as long as they collaborate with other soil microbes. It was experimentally proven that mechanisms underlying the increased P-uptake in arbuscular mycorrhizal plants were solely due to AMF synergistic interactions with P-solubilizing microorganisms and/or greater soil volume explored by the AMF hyphae (Antunes et al., 2007). Otherwise, the AMF unlike ectomycorrhiza are not able to neither solubilize phosphate nor decompose organic matter (Simard and Austin, 2010). The well-known activities of nitrogen-fixing bacteria and P-solubilizing microorganisms improving the bioavailability of the major plant nutrients N and P are very much enhanced in the mycorrhizosphere where synergistic interactions of such microorganisms with mycorrhizal fungi have been demonstrated (Barea et al., 2002). In particular, mycorrhizal inoculation improved the establishment of both inoculated and indigenous P-solubilizing rhizobacteria and, again P-solubilizing rhizobacteria usually behave as mycorrhiza-helper-bacteria, promoting mycorrhiza establishment by both the indigenous and the inoculated mycorrhizal fungi (Barea et al., 2002).

Arbuscular mycorrhizal fungi also interact with decomposer fungi (Soka and Ritchie, 2014) and phosphate solubilizing fungi (PSF) synergistically (Osoria and Habte, 2001). Accordingly, presence of mycorrhizal fungi is known to alter the rates of above and below ground litter decomposition due to chemical changes in the roots and interactions with the decomposer fungi (Soka and Ritchie, 2014). PSF were also observed to have lesser effect in plant nutrition when applied alone and maximum effect took place when both AMF and PSF were inoculated showing the synergistic interaction between the AMF and PSF (Osoria and Habte, 2001). In the meantime, AMF are also known to have antagonistic relationship with root pathogens (Soka and Ritchie, 2014) and even leaf pathogens (Parniske, 2008).

Arbuscular mycorrhizal fungi may also interact with each other synergistically. It was experimentally found out that AMF effects are greater when AMF consortia inoculums are applied than single AMF (Banerjee et al., 2013). After long years of observation, Barea et al.,. (2011)concluded that the use of native AMF consortia has the maximum effect. A meta-analysis on 306 studies also indicated that plant response was substantially lower when plants were inoculated with single AMF species, compared with inoculations with multiple AMF species (Hoeksema et al., 2010). This could be due to synergistic interaction between the various AMF species. Different species of AMF have different hyphal growth patterns, anastomoses and branching frequencies (Parniske, 2008). These differences probably reflect different strategies and the occupation of different niches within the soil (Parniske, 2008).

AMF AND MEASURABLE RESTORATION ATTRIBUTES

Improved plant fitness (survival, growth and reproduction), nutrient uptake and accumulation, tolerance of adverse conditions (biotic and abiotic stresses) and altering plant community structure [competition/facilitation, diversity (richness and evenness) and succession] and that of animal communities (Direct effects on organisms which feed on fungi and indirect effects due to changes in plant fitness) were identified to be the pivotal role AMF play in ecological restoration (Brundrett and Abbott, 2002).

Based on Aronson et al.,. (1993), the functional and structural attributes to measure ecological restoration include; soil organic matter, soil water relation, nutrient cycling index, plant diversity and soil microbial diversity and abundance, and plant productivity. Therefore, the role of AMF in soil organic matter content, soil water relation, nutrient cycling index, plant stress tolerance, plants survival, establishment and growth on degraded soils, plant diversity, soil microbial diversity and abundance, and plant succession, competition/facilitation and productivity is highlighted below.

AMF Improve Soil Aggregation; Hence Increase Soil Organic Matter and Soil Water Relation

Fungi and most importantly AMF may be the most effective soil organisms in stabilizing soil structure (Augé, 2004). AMF hyphae grow into the soil matrix to create the skeletal structure that holds primary soil particles together to form soil aggregates (Augé, 2004; Al-Karaki, 2013).

AMF also improve soil aggregation by influencing bacterial communities that can improve soil aggregate formation (Rilling, 2004). Furthermore, the dead AMF hyphae produce glomalin which is hydrophobic stable aggregate former (Barea et al., 2002; Simard and Austin, 2010). Hence, AMF increase both soil aggregation and stability. AMF may stabilize soils up to 5 months after their host's death (Soka and Ritchie, 2014).

Meanwhile, as a result of the significant amount of mycorrhiza derived soil carbon (Rilling, 2004) and improved soil aggregation and stability, AMF increase soil organic matter content and stability (Rilling, 2004; Leifheit et al., 2014). Improved soil aggregation also increases soil water relation. It was observed that a naturally non-mycorrhizal plant planted in mycorrhizal soils tolerated drought more than the ones planted in a non-mycorrhizal soils indicating that AMF hyphae improves water holding capacity of soils (Marschner, 1995).

AMF Improve Plant Nutrition and Nutrients Cycling Index

The most important role of AMF is their role in phosphorous nutrition (Skujins and Allen, 1986). There are also data indicating that AMF can transfer nitrogen from one plant to another (e.g., Requena et al., 2001), increase the utilization of different forms of nitrogen by plants and can also take up nitrogen directly and transfer it to host roots (Govindarajulu et al., 2005). However, there is considerable doubt as to the cost-benefit of AMF in plant N nutrition (Smith and Smith, 2011). Although few data exist, AMF were observed to improve potassium nutrition in plants (Dag et al., 2009; Garcia and Zimmermann, 2014). AMF can also increase the uptake of other macro and micro nutrients by plants (Birhane et al., 2012). Generally, the external mycelium of AMF establishes an underground network that links the different plants and hence sequester carbon, nitrogen, and phosphorous and also allow the transfer of these nutrients among plants (Rodriguez-Echeverria et al., 2007). These important roles of AMF therefore play great role in nutrient cycling where the need for further nutrient inputs is significantly reduced (Gianinazzi et al., 2010; Al-Karaki, 2013).

Arbuscular mycorrhizal fungi not only improve nutrient cycling but also reduce nutrient leaching from the soil (Rodriguez-Echeverria et al., 2007). In a comprehensive assessment done by Bender et al.,. (2015), it was possible to determine the role AMF have in nutrient cycling and leaching. Accordingly, it was determined that while AMF inoculation increased nutrient uptake by plants it also reduced leaching of dissolved organic N and un-reactive P (Bender et al., 2015).

AMF Increase Plants' Abiotic Stress Tolerance

It was, several times, demonstrated that AMF can increase plants' tolerance to drought and salinity (Al-Karaki, 2013). AMF are also known to alleviate heavy metal stress in plants (Leyval et al., 1997; Hildebrandt et al., 2007; Soares and Siqueira, 2008; Amir et al., 2013). By inoculating plants with drought tolerant AMF, up to 42% reduction in plants' water requirement could be achieved (Gianinazzi et al., 2010). Also, Navarro et al.,. (2013) found out that, *Citrus* rootstocks inoculated with AMF showed significantly increased growth than non-inoculated individuals despite the fact that inoculated individuals were irrigated with saline water and the non-inoculated ones got irrigated with non-saline water.

The mechanism by which AMF increase plants' tolerance to drought, salinity and heavy metal stresses is mainly nutritional (Marschner, 1995; Soares and Siqueira, 2008; Birhane et al., 2012; Al-Karaki, 2013; Navarro et al., 2013). Soares and Siqueira (2008) demonstrated that both P fertilization and AMF inoculation of plants significantly improved plants' growth on heavy metal polluted soils. Hence, they concluded, AMF increase plants' heavy metal stress tolerance mainly through P nutrition.

The non-nutritional mechanisms by which AMF increase plants' tolerance to drought include; hormonal changes, hyphal soil improvement (delayed soil drying), hyphal ability to scavenge water from micro-pores, increased plants' photosynthetic rate, and accumulation of compatible osmolites (Marschner, 1995; Birhane et al., 2012; Al-Karaki, 2013). Likewise, immobilizing heavy metals in their biomass mainly cell wall, vesicles and in the glomaline is the non-nutritional mechanism by which AMF improve plants' tolerance to heavy metals stress (Hildebrandt et al., 2007).

The positive AMF effects on plants' drought tolerance can improve plants' salinity tolerance as well. Better water intake by plants can effectively dilute salts within the plants' cells (Larcher, 1995). Other non-nutritional mechanisms by which AMF improve plants' salinity tolerance include; exclusion of salt from plant cells by accumulating the salt within the fungal hyphae, production of enzymes involved in antioxidant defense, and change in cell wall elasticity and membrane stability (Al-Karaki, 2013).

AMF Increases Plants' Resistance and Tolerance to Pathogens and Herbivores

There are several published articles showing the role of AMF in increasing plant tolerance against biotic stressors. The meta-analysis of 144 published papers clearly reveals that (Yang et al., 2014). Considering the role AMF have in bioprotection, Gianinazzi et al.,. (2010) described AMF as 'health insurance' of plants. One mechanism by which AMF increase plants' pathogen tolerance could be the synergistic interaction of AMF have with plant growth promoting rhizobacteria (PGPR). PGPR have a very well documented role in plant pathogen inhibition (Figueiredo et al., 2010). The fact that AMF stimulate the synthesis of plant secondary metabolites (Gianinazzi et al., 2010) may also explain why AMF inhibit herbivory. Plants' secondary metabolites are known to have role in plants' defense against herbivores (Larcher, 1995). The other reason by which AMF increase plants' herbivory tolerance is compensatory growth. A microcosm investigation revealed that mycorrhizal plants did not show a reduction in total above ground biomass despite their leaves being fed by grasshoppers indicating that mycorrhiza helped the plant to compensate in growth after herbivory (Kula et al., 2005).

AMF Increase Tree/Shrub Seedlings Growth, Productivity, Field Survival and Establishment on Degraded Lands

Lekberg and Koide (2005), carried out a meta-analysis based on 290 published experiments to determine the role of AMF on plant growth and productivity. The analysis also determined the effects of three common AMF management methods; inoculation, short fallow, and reduced soil disturbance. The result of the meta-analysis revealed that AMF generally increase individual plant's growth and productivity. Inoculation and short fallow resulted in significantly positive effects on plants' growth and productivity (Lekberg and Koide, 2005). A recent meta-analysis on

304 papers also concluded that AMF inoculation increases the growth and productivity of plants grown alone (Lin et al., 2015). A similar result was also reported by Birhane et al.,. (2014). Huante et al.,. (2012) also did experiment on six tree species and found out that AMF inoculation has significant effect on seedlings growth and most significantly slow growing tree species. Figure 1 below shows how AMF inoculation can significantly increase tree seedlings growth.

Figure 1: ***Acacia koa*** **A. Gray grew significantly tall in a low-P soil when inoculated with AM fungus (adopted from Miyasaka et al., 2003).**

Tree survival and field establishment is an important factor in the restoration of degraded lands. Hence, AMF are important since they can significantlyimprovetreeseedlingsfieldsurvivalandestablishment. Pouyu-Rojas and Siqueira (2000), Habte et al.,. (2001), Ouahmane et al.,. (2006), Dag et al.,. (2009), Kapulnik et al.,. (2010), Karthikeyan and Krishnakumar (2012), and Manaut et al.,. (2015) have demonstrated the positive effect AMF have in these regards. Pouyu-Rojas and Siqueira (2000) investigated the AMF effect on seven tree species seedlings survival and establishment on degraded pot soils. They found out that AMF inoculation in the nursery or during transplanting have equally significantly positive effect on trees survival and establishment. Later on Habte et al.,. (2001) determined the effect AMF nursery inoculation has on field establishment of *Acacia koa* and accordingly, AMF was shown to improve transplanted tree seedlings growth and establishment by increasing seedlings P nutrition. The role of native AMF inoculation was also demonstrated to have a significant positive effect on the field survival and establishment of *Cupressus atlantica* Gaussen seedlings on a degraded Moroccan field site (Ouahmane et al., 2006).

Similarly, Kapulnik et al.,. (2010) determined AMF nursery inoculation effect on seedlings field establishment and growth of *Olea europaea* L. Meanwhile, they were able to observe that AMF inoculation improved seedlings field performance significantly and most importantly for the first 2.5 years from transplanting. They also observed that AMF effect decreased with increasing seedlings age. Karthikeyan and Krishnakumar (2012) also determined AMF effect on survival and establishment of *Eucalyptus tereticornis* Sm. on pot soil of highly degraded origin (mine spoils). Meanwhile, they were able to observe that AMF inoculation almost doubled seedling survival and significantly increased establishment. Recently, Manaut et al.,. (2015) demonstrated that native AMF consortia inoculation of *Ceratonia siliqua* L. seedlings more than doubled seedlings' survival and significantly improved seedlings' height and collar diameter.

AMF Drive Succession and Influence Plant Community Structure

According to Janos (1980), the mycorrhizal fungus status and the fertility of soil influence the occurrence of plant species. It is also hypothesized that AMF are drivers and as well, passengers of plant community succession (Zobel and pik, 2014). Meanwhile, the AMF status of a site determines the composition of a seral plant community, and the composition of that particular seral plant community determines the composition of infective AMF communities which will further influence the composition of the next seral plant community (Janos, 1980; Renker et al., 2004). Thus, if specific compatible relationships between certain AMF and plant taxa are required for mutual symbiont survival, the loss of compatible AMF species or individuals may limit the distribution of a particular plant species (Renker et al., 2004). Plant-soil feedback (plant-AMF feedback) is also an important concept explaining the role AMF have in succession (Kikvidze et al., 2010). Positive feedbacks promote the development of early successional communities and negative feedbacks promote plant species replacement to drive succession (Kikvidze et al., 2010).

Arbuscular mycorrhizal fungi could also potentially influence plant community structure by affecting richness or evenness of coexisting plants (Brundrett and Abbott, 2002). Only some 240 AMF morphospecies have been described forming associations with 80% of terrestrial plants (Lee et al., 2013). This indicates that AMF have no host specificness. Meanwhile, a single mycorrhizal fungus can link different plants together, thus forming mycorrhizal networks (Simard and Austin, 2010; Song et al., 2014). These networks have been shown to facilitate regeneration of new seedlings, alter species interactions, and change the dynamics of plant communities therefore, increasing plant diversity (Simard and Austin, 2010). Sowing seeds of plant species in microcosms that resembled the grassland community of the temperate zone, on soils of AMF inoculated and non-inoculated, Van der Heijden et al.,. (1998), Vogelsang et al.,. (2006), and Schnitzer et al.,. (2011) were able to observe that AMF inoculation improved plant community diversity by mainly increasing plants' fitness and evenness. AMF may also be important organisms to inhibit invasion by alien species. This could be indirectly by reducing the vacant niche through increased native plants survival, establishment and diversity or can be by direct inhibition of invasives. Janos et al.,. (2013) reported that the presence of established extraradical mycelium prevented the survival and establishment of seedlings migrating from another ecosystem.

AMF Increase Soil Microbial Diversity and Abundance

Arbuscular mycorrhizal fungi hyphae and root litter are the most abundant carbon source in the soil (Brundrett and Abbott, 2002). Therefore, AMF provide increased supply of energy for soil microbes to flourish. The fact that AMF influence plant communities is also considered to be one of the potential mechanisms by which AMF influence soil microbial communities (Rilling, 2004). Furthermore, AMF hyphal exudates may also stimulate microorganisms present in the mycorrhizal hyphosphere. However, the effect is variable: AMF hyphal exudates may stimulate some microorganisms but still inhibit others (Herman et al., 2012). Hence, AMF may increase the diversity and abundance of microorganisms that are beneficial to plants' growth and health.

AMF Effects on Plant Community Primary Productivity and Plant Competition

Arbuscular mycorrhizal fungi inoculation increases plant productivity at community level and the effect increases with increase in plant species richness following the common ascending but asymptotic diversity-productivity pattern (Schnitzer et al., 2011). At low plant diversity soil microbes suppress plant productivity since their pathogenic effect increases and as the plant diversity increases, plant productivity can increase up to fivefold (Figure 2; circle and triangle). In the absence of soil microbes plant productivity has a weak positive linear relationship with plant diversity (Figure 2; square).

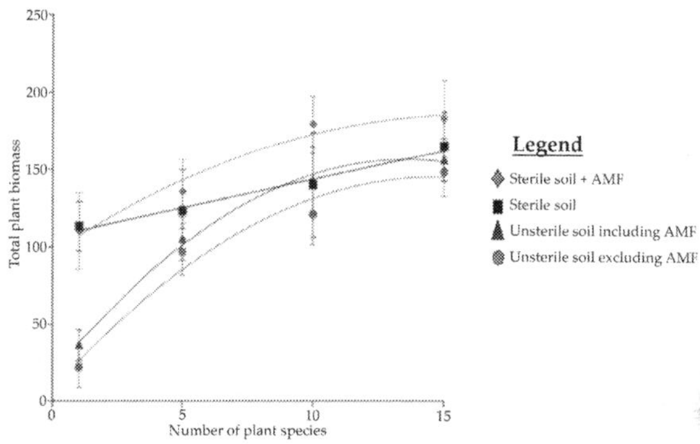

Figure 2: The role of microbes and AMF in plant diversity-productivity relationship (adopted from **Schnitzer et al., 2011**).

Sowing seeds of plant species that resembled the grassland community of the temperate zone on AMF inoculated and non-inoculated soils, Van der Heijden et al.,. (1998) and Vogelsang et al.,. (2006) were able to determine AMF effect on plant productivity at community level. Accordingly, AMF improved plant community productivity (Van der Heijden et al., 1998; Vogelsang et al., 2006). Furthermore, Van der Heijden et al.,. (1998) observed that increasing AMF richness resulted in increased productivity while Vogelsang et al.,. (2006) observed although AMF species richness increased productivity, the effect was not significant compared to single AMF inoculums effect. So, the former observation is Van der Heijden et al.,. (1999) argued, due to niche complementarity while the latter is Vogelsang et al.,. (2006) argued, due to sampling effect. The "niche complementarity" theory argues that the presence of many species and functional types results in more complete utilization of resources because different species specialize on different resources, resulting in higher overall productivity while the "sampling effect" theory argues species identity is more important than diversity and asserts that productivity increases with diversity solely due to an increased probability that communities with more species contain a few very productive species that disproportionately contribute to community-wide productivity (Schnitzer et al., 2011).

Despite the fact that the above research observations reveal that AMF increase plant productivity at community level, the recent meta-analysis based on 304 study results, which also cited the above research observation, found out that, at community level, AMF inoculation either has no effect on plant productivity or even has a negative effect (Lin et al., 2015). AMF inoculation increases plant productivity at community level only when experiments were conducted in the green house (Lin et al., 2015). Klironomos et al.,. (2000) were, in a greenhouse setting, able to determine the AMF effect on plant diversity and that of productivity. Accordingly, it was observed that AMF inoculation increased plant productivity but not for all AMF species. Inoculating *Glomus intraradices* N. C. Schenck and G. S. Sm. even lowered productivity compared to the non-inoculated plant community (Klironomos et al., 2000).

Similarly, variable AMF effect is observed in plant competition. Plant species competitive ability response to AMF inoculation depends on plants' functional group, mycorrhizal status, plants' life history (Scheublin et al., 2007; Lin et al., 2015), and also maybe below ground functional traits of the plant species (Birhane et al., 2014). The meta-analysis conducted by Lin et al.,. (2015) also concludes, AMF inoculation significantly increases N-fixing forbs, decreases C3 grasses and non-N-fixing forbs and woody plants, and has no effect on C4 grasses competitive ability whether these functional groups compete intra or inter-specifically (Lin et al., 2015). According to Birhane et al.,. (2014), in a pot experiment, AMF inoculation did not have positive effect on the competitive ability of both *Acacia etbaica* Schweinf. and *Boswellia papyrifera* Hochst. seedlings grown together. In other instances, mycorrhizal networks may result in asymmetric competition by favoring strong carbon-donor roots (Weremijewicz and Janos, 2013) or vice-versa (Walder et al., 2012).

AMF BIOTECHNOLOGY FOR THE RESTORATION OF DEGRADED LANDS

Degraded lands have low level of infective AMF and nursery seedlings around degraded sites may less likely be infected with sufficient AMF (e.g., Michelsen, 1992). Therefore, these sites can support the growth of late successional tree species when appropriate AMF inocula are reintroduced. Late successional tree species are obligately mycotrophic and may necessarily require AMF for their survival and fitness (Janos, 1980). More importantly, at the early stages of seedlings growth, mycorrhizal early/mid successional tree/shrub species can be even more AMF dependent than the late successional ones (Kiers et al., 2000). Therefore, AMF inoculation could potentially be considered as an important biotechnological tool in degraded lands restoration.

Arbuscular mycorrhizal fungi show no host specificity to forge symbiotic relationship with plants and are very ubiquitous, found almost in every soil (Abbott and Robson, 1991; Brundrett and Abbott, 2002; Barea et al., 2011; Al-Karaki, 2013). Hence, many researchers argue that AMF inoculation is likely to be valuable in only few conditions such as mine fields where indigenous AMF inoculum is surely little or none available (Brundrett and Abbott, 2002). Koide and Mosse (2004) suggested instead of going for AMF inoculation it would be quite economical and appropriate to focus on managing the indigenous AMF population of a site. According to Renker et al.,. (2004), inoculation is an important but the last option. However, contrary to having several dispersal agents such as; wind, water, rodents, birds, worms, and ants (Brundrett and Abbott, 2002), AMF were observed to have poor dispersal. Accordingly, Hailemariam et al.,. (2013) were able to observe that within a single piece of farm land, soil AMF status and infectiveness can vary in short distances indicating poor dispersal. Similarly, Friese and Allen (1991), also indicated that AMF have poor dispersal. Therefore, to overcome the dispersal limitation of AMF, inoculation may be a worthily intervention.

Meanwhile, AMF inoculation has proved to be effective under wide range of soil conditions (Janos, 1980; Brundrett and Abbott, 2002) including on soils with good AMF abundance (e.g., Banerjee et al., 2013). Positive AMF effect is not ensured by the presence of abundant indigenous AMF but by both abundance (quantity) and efficiency (quality) of indigenous fungal populations (Onguene and Kuyper, 2005). Veiga et al.,. (2011) also demonstrated that AMF inoculation suppressed weeds and, interestingly enough, hypothesized that AMF inoculation could suppress ruderal plants which are known to invade degraded sites (Veiga et al., 2011). This is particularly important in ecological restoration since ruderal plants could invade degraded lands and compete with tree/shrub seedlings planted.

If the importance of AMF inoculation in the restoration of degraded lands is agreed, the next question to ask will be; what kind of inocula should be prepared? AMF show wide range of functional diversity (Johnson et al., 1997; Klironomos, 2003; Smith et al., 2011) and their effect is within the mutualism-parasitism continuum (Johnson et al., 1997). Likewise, Hoeksema et al.,. (2010) summarized that certain plants functional groups viz. non-N-fixing forbs and woody plants and C4 grasses show more positive responses to AMF inoculation. Klironomos (2003) also demonstrated that exotic-native AMF strain-host or vice-versa combination results in highly parasitic interaction. Therefore, deciding on the type of inoculum to prepare is a very important step. Based on the currently available data, the use of native inocula should be preferred to the use of exotic inocula. Early seral AMF should be used when seedlings are inoculated for restoration, even for late seral tree species (Allen et al., 2003). Late seral AMF have big spores and demand much carbon and hence, seedlings may not benefit from them. Instead, seedlings benefit from early successional AMF which are usually having small spores and smaller carbon demand (Allen et al., 2003). Likewise, the use of inocula from grasslands is promoted. AMF abundance in grasslands can be more than tenfold than that of in the forestlands and AMF from grasslands do have significantly high inoculation effect (Fischer et al., 1994). That was why Onguene and Kuyper (2005)applied fresh grassland whole-soil inoculum on various soils and three tree species seedlings. According to the result Onguene and Kuyper (2005) obtained, although early successional grassland inoculum had positive effect for most of the cases (80%), the fact that it is an inoculum from grassland resulted in significantly negative effect on *Terminalia superba* Engl. and Diels seedlings grown on agricultural and early successional forest soils. Hence, Onguene and Kuyper (2005) concluded; allochthonous AM inocula may not be always effective. Hence, the use of planting site adapted AM inocula may be recommended. The other reason for the observed negative effect may also be related to host plant's fungi preference (Onguene and Kuyper, 2005).

There are data to demonstrate that inocula from conspecific source show better affinity to the host plants' root (e.g., Kiers et al., 2000). Similarly, there are data to show that plant species even that do co-occur may prefer to associate with distinct AM fungi communities (e.g., Wubet et al., 2006; Davison et al., 2011). There are also data to show that distinctively different AMF communities colonize seedlings' and adults' roots of a single tree species (e.g., Wubet et al., 2009). Therefore, one has to ask; does inoculating seedlings with AM inocula from seedlings' rhizosphere or adults' deliver better positive effect? Kiers et al.,. (2000) have found out that although conspecific inocula from adults had better affinity to inoculated seedlings, the effect on their growth was mostly relatively small showing that, inocula even from conspecific adults, may not be suited for seedlings inoculation.

Selecting few of the dominant planting site adapted AMF species, multiplying them and applying as inocula may not be also a very good idea specially when there are established AMF in the planting site. Increasing the density of few of the dominant AMF species and applying as inocula had resulted in negative effects on plant growth by disrupting indigenous AMF community structure and thereby creating competition among AMF to ultimately result in inoculum failure (Janoušková et al., 2013). Therefore, in areas with low levels of indigenous AMF abundance, multiplying all not only the dominant AMF species and applying all may be the best option.

The AMF richness in AM inocula is considered to improve inocula effectiveness. Plant response is substantially lower when inoculated with single AMF species and the response keeps increasing from multiple fungal species to whole-soil inoculums (Hoeksema et al., 2010). Likewise, Barea et al.,. (2011) compiling long years of experience in AMF research recommend the use of autochthonous foundation shrub inoculated with autochthonous AMF consortia inoculums to best restore degraded lands of the Mediterranean. The shrub not only acts as a foundation species but also serves as a resource island for AMF (Barea et al., 2011). However, not all ecologists agree by the application of AMF species rich inocula; some argue that better results due to inocula with better AMF species richness is due to sampling effect and selecting single effective AMF species should get the attention of restoration ecologists. Sampling effect is discussed earlier.

The other challenge associated with AMF biotechnology is related with inocula production for large-scale application. This is due mainly to the obligate nature of AMF. Meanwhile, AMF cannot be cultured axenically (Azcón-Aguilar et al., 1999; Fortin et al., 2005) and host plant based AMF multiplication is mandatory. These host plant based conventional inocula production methods (substrate based pot culturing and substrate free methods of hydroponics and aeroponics techniques) are costly and large scale production of AMF inocula may hardly be possible. Effective monexenic *in vitro* culturing of AMF has been made possible few decades ago (Bécard and Fortin, 1988) and in India, using this method, large-scale industrial production of biologically clean AMF inocula was possible (Adholeya et al., 2005). Readers are directed to read Adholeya et al.,. (2005) and Cranenbrouck et al.,. (2005) to grasp the potential and the technique of monoxenic *in vitro* AMF culture production for large-scale application. Readers are also directed to read Azcón-Aguilar et al.,. (1999) to get proper definitions of axenic and monoxenic cultures.

However, until now, monexenic *in vitro* culturing is not widely practiced. This is due mainly to the fact that; (1) undesired contamination is hardly avoidable and the technique is technology and skill demanding (Bago and Cano, 2005), (2) there are ethical and legal concerns, and (3) it is rather very hard to identify each genotype (even morphotype) hence, most if not all, AMF are not readily culturable (Fortin et al., 2005). AMF momoxenic *in vitro* culturing uses transformed [using *Rhizobium rhizogenes* (Riker et al.,.) Young et al.,.] hairy roots as host owing to the fact that these hairy roots are better suited than the non-transformed hairy roots since they grow on hormone free media and without developing shoots and leaves (Puri and Adholeya, 2013). Meanwhile, AMF monoxenic culture as it is practiced now could potentially be challenged with biosafety related issues.

Due to the lack of cheap and easy AMF inocula production for large scale application, managing the *in situ* AMF is sometimes considered to be an effective AMF biotechnology for the restoration of degraded lands. The meta-analysis by Lekberg and Koide (2005) showed that short fallow could be as good as inoculation to improve plants growth and productivity. It was shown that an obligately arbuscular mycorrhizal pioneer nurse shrub *Lavandula stoechas* L. improved the field survival and establishment of *Cupressus atlantica* Gaussen seedlings by increasing, among others, *in situ* infective AMF abundance (Duponnois et al., 2011). Kumar et al.,. (2010) also compared different plant composition effects on *in situ* management of AMF on a degraded coal mine spoil. Accordingly, they demonstrated that using cover crops mainly grasses and N-fixing shrubs in the plant composition, significantly enhanced AMF abundance, diversity and infectiveness. Hence, AMF can be manipulated by fallowing or/and by designing the plant species composition to ultimately result in increased AMF abundance which intern facilitates restoration. However, some investigations indicated that grass cover can significantly suppress individual tree/shrub seedlings-saplings growth (Riginos, 2009) or may have variable seasonal effects (Good et al., 2014). Therefore, investigation on cover plant management options to effectively manage AMF and facilitate tree/shrub seedlings growth can be an important research topic.

Nowadays, substrate free inocula preparation methods and *in vitro* production on excised plant roots are being intensively researched to make AMF inoculation less costly (Ijdo et al., 2011). The pot culture inocula preparation method, although it is labor intensive and costly, can be a source of employment especially in developing countries. Therefore, pot culture based AMF biotechnology will remain to be a feasible way of degraded lands restoration in most parts of the world. Figure 3 shows the simplified schematic model of degraded lands restoration using AMF.

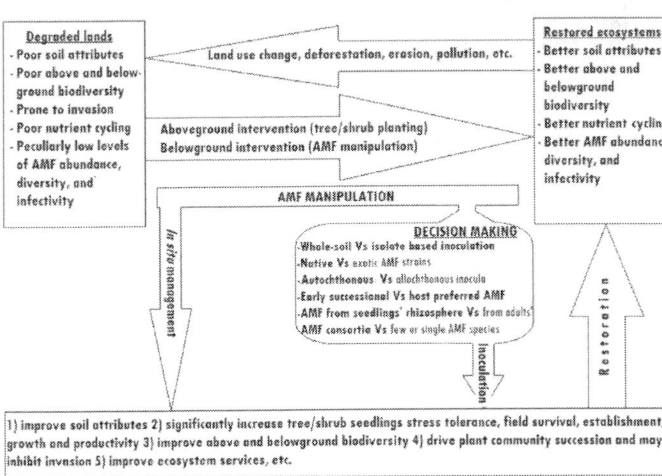

Figure 3: Simplified schematic model showing degraded lands restoration

through AMF manipulation.

CONCLUSION

This review paper has compiled facts regarding the AMF role in the above and belowground ecosystem processes relevant to ecological restoration. Accordingly, it is possible to conclude that AMF; have a well documented positive role in nutrient cycling and improved soil attributes. AMF also improve plants' tolerance to biotic and abiotic stresses, and significantly increase tree/shrub seedlings survival, establishment and growth. AMF play pivotal role in plant community succession and may directly or indirectly prevent invasion by alien plant species. At plant community level, AMF increase both above and below ground biodiversity but their effect on primary productivity maybe low. The AMF effect on plant competition is also variable and mostly negative. Available data as of yet, indicate that there are very few outfield experiments done on AMF effects on tree/shrubs seedlings survival and establishment. This review was not also able to clearly trace a research result showing the AMF effect on the competitive ability of tree seedlings planted with annual and perennial grass and/or herbaceous weeds. Based on the currently available data, however, it can be concluded that AMF inoculation can significantly increase the success of degraded lands restoration and for better results reducing competitors and seedlings density (increased seedling spacing) is recommended.

Based on the data reviewed in this article, we recommend for future AMF effect researches to give emphasis to outfield experiments. The AMF effect on the competitive ability of tree seedlings compared with annual and perennial herbaceous weeds should be investigated. Data reviewed here showed that almost all research observations conducted on AMF effect at community level are on microcosms of grasslands; and mainly temperate grasslands. Future researches should focus on forest communities of both the temperate and tropics. For an effective large scale application of AMF inocula biotechnology, pot based inocula multiplication will remain to be significantly cost ineffective. Therefore, investigating and researching on cost effective multiplication methods of substrate free and *in vitro* culture and/or optimization of the effects of low-cost fresh AMF inoculation techniques like using grassland top soil or managing AMF *in situ* using several cover crops including grasses need further attention in the future. Optimization of monoxenic *in vitro* AMF culture products and using non-transformed hairy root organ could also be an important research area until axenic *in vitro* AMF culturing is ultimately made possible.

AUTHOR CONTRIBUTIONS

FA did all the data gathering and write-up. TB and EB considerably contributed intellectually by providing comments and guidance at every milestone of the manuscript development. All authors approved publication of the article.

ACKNOWLEDGMENTS

The authors would like to thank Dr. Marc Ducousso and Dr. Laila Pamela Partida-Martinez for their valuable and enriching comments. Reviewers of the first draft are also acknowledged.

FOOTNOTES

[1]https://www.cbd.int/sp/targets/

[2]http://www.wri.org/faq/what-degraded-land

REFERENCES

1. Abbott L. K., Robson A. D. (1991). Factors influencing the occurrence of vesicular-arbuscular mycorrhizas. *Agric. Ecosyst. Environ.* 35121–150. 10.1016/0167-8809(91)90048-3
2. Adholeya A., Tiwari P., Singh R. (2005). "Large scale inoculum production of arbuscular mycorrhizal fungi on root organs and inoculation strategies," in *Soil Biology In Vitro Culture of Mycorrhizae* Vol. 4 eds Declerck S., Strullu D. G., Fortin A., editors. (Berlin: Springer-Verlag;), 315–338.

3. Aerts R., Honnay O. (2011). Forest restoration, biodiversity and ecosystem functioning. *BMC Ecol.* 11:29 10.1186/1472-6785-11-29 [PMC free article] [PubMed]

4. Alexander I., Ahmad N., See L. S. (1992). The role of mycorrhizas in the regeneration of some Malaysian forest trees. *Philos. Trans. Biol. Sci.* 335 379–388. 10.1098/rstb.1992.0029

5. Al-Karaki G. N. (2013). "The role of mycorrhiza in the reclamation of degraded lands in arid environments," in *Developments in Soil Classification, Land Use Planning and Policy Implications: Innovative Thinking of Soil Inventory for Land Use Planning and Management of Land Resources*, eds Shahid S. A., Taha F. K., Abdelfattah M. A., editors. (Dordrecht: Springer Science+Business Media;), 823–836.

6. Allen E. B., Allen M. F., Egerton-Warburton L., Corkidi L., Gomez-Pompa A. (2003). Impacts of early and late-seral mycorrhizae during restoration in seasonal tropical forest, Mexico. *Ecol. Appl.* 13 1701–1717. 10.1890/02-5309

7. Amir H., Lagrange A., Hassaïne N., Cavaloc Y. (2013). Arbuscular mycorrhizal fungi from New Caledonian ultramafic soils improve tolerance to nickel of endemic plant species. *Mycorrhiza* 23 585–595. 10.1007/s00572-013-0499-6

8. Angelini C., Altieri A. H., Silliman B. R., Bertness M. D. (2011). Interactions among foundation species and their consequences for community brganization, biodiversity, and conservation. *BioScience* 61 782–789. 10.1525/bio.2011.61.10.8

9. Antunes P. M., Schneider K., Hillis D., Klironomos J. N. (2007). Can the arbuscular mycorrhizal fungus *Glomus intraradices* actively mobilize P from rock phosphates? *Pedobiologia* 51 281–286. 10.1016/j.pedobi.2007.04.007

10. Aradottir A. L., Hagen D. (2013). Ecological restoration: approaches and impacts on vegetation, soils and society. *Advan. Agron.* 120173–222. 10.1016/b978-0-12-407686-0.00003-8

11. Aronson J., Fled C., Le Floc'h E., Ode C., Pontanier R. (1993). Restoration and rehabilitation of degraded ecosystems in arid and semi-arid lands: a view from the south. *Restor. Ecol.* 1 8–17. 10.1111/j.1526-100X.1993.tb00004.x

12. Augé R. M. (2004). Arbuscular mycorrhizae and soil/plant water relations. *Can. J. Soil Sci.* 84 373–381. 10.4141/S04-002

13. Azcón-Aguilar C., Bago B., Barea J. M. (1999). "Saprophytic growth of arbuscular mycorrhizal fungi," in *Mycorrhiza: Structure, Function, Molecular Biology and Biotechnology*, eds Varma A., Hock B., editors. (New York, NY: Springer;), 391–407.

14. Bago B., Cano C. (2005). "," in *Soil Biology In Vitro Culture of Mycorrhizae* Vol. 4 eds Declerck S., Strullu D. G., Fortin A., editors. (Berlin: Springer-Verlag;), 111–138.

15. Bai Z. G., Dent D. L., Olsson L., Schaepman M. E. (2008). *Soil Use Manag.* 24 223–234. 10.1111/j.1475-2743.2008.00169.x

16. Banerjee K., Gadani M. H., Srivastava K. K., Verma N., Jasrai Y. T., Jain N. K. (2013). Screening of efficient arbuscular mycorrhizal fungi for Azadirachta indica under nursery condition: a step towards afforestation of semi-arid region of Western India. *Braz. J. Microbiol.* 44 587–593. 10.1590/S1517-83822013005000046

17. Barea J. M., Azcon R., Azcón-Aguilar C. (2002). Mycorrhizosphere interactions to improve plant fitness and soil quality. *Antonie Van Leeuwenhoek* 81 343–351. 10.1023/A:1020588701325

18. Barea J. M., Palenzuela J., Cornejo P., Sánchez-Castro I., Navarro-Fernández C., Lopéz- García A., et al.,. (2011). Ecological and functional roles of mycorrhizas in semi-arid ecosystems of Southeast Spain. *J. Arid Environ.* 75 1292–1301. 10.1016/j.jaridenv.2011.06.001

19. Bécard G., Fortin J. A. (1988). Early events of vesicular-arbuscular mycorrhiza formation in RiT-DNA transformed roots. *New Phytol.* 108 211–218. 10.1111/j.1469-8137.1988.tb03698.x

20. Bender S. F., Conen F., Van der Heijden M. G. A. (2015). Mycorrhizal effects on nutrient cycling, nutrient leaching and N_2O production in experimental grassland. *Soil Biol. Biochem.* 80 283–292. 10.1016/j.soilbio.2014.10.016

21. Bever J. D., Richardson S. C., Lawrence B. M., Holmes J., Watson M. (2009). Preferential allocation to beneficial symbiont with spatial structure maintains mycorrhizal mutualism. *Ecol. Lett.* 12 13–21. 10.1111/j.1461-0248.2008.01254.x

22. Birhane E., Kuyper T. W., Sterck F. J., Bongers F. (2010). Arbuscular

mycorrhizal associations in *Boswellia papyrifera* (frankincense-tree) dominated dry deciduous woodlands of Northern Ethiopia. *For. Ecol. Manag.* 260 2160–2169. 10.1016/j.foreco.2010.09.010

23. Birhane E., Kuyper T. W., Sterck F. J., Gebrehiwot K., Bongers F. (2015). Arbuscular mycorrhiza and water and nutrient supply differently impact seedling performance of acquisitive and conservative dry woodland species. *Plant Ecol. Divr.* 8 1–13. 10.1080/17550874.2014.992488
24. Birhane E., Sterck F. J., Bongers F., Kuyper T. W. (2014). Arbuscular mycorrhizal impacts on competitive interactions between *Acacia etbaica* and *Boswellia papyrifera* seedlings under drought stress. *J. Plant Ecol.* 1 298–308. 10.1093/jpe/rtt031
25. Birhane E., Sterck F. J., Fetene M., Bongers F., Kuyper T. W. (2012). Arbuscular mycorrhizal fungi enhance photosynthesis, water use efficiency, and growth of frankincense seedlings under pulsed water availability conditions. *Oecologia* 169 895–904. 10.1007/s00442-012-2258-3
26. Brundrett M. C. (2009). Mycorrhizal associations and other means of nutrition of vascular plants: understanding the global diversity of host plants by resolving conflicting information and developing reliable means of diagnosis. *Plant Soil* 320 37–77. 10.1007/s11104-008-9877-9
27. Brundrett M. C., Abbott L. K. (2002). "Arbuscula mycorrhiza in plant communities," in *Plant Conservation and Biodiversity*, eds Sivasithamparam K., Dixon K. W., Barrett R. L., editors. (Dordrecht: Kluwer Academic Publishers;), 151–193.
28. Brundrett M. C., Bougher N., Dell B., Grove T., Malajczuk N. (1996). *Working with Mycorrhizas in Forestry and Agriculture, ACIAR Monograph 32.* Canberra: Australian Centre for International Agricultural Research.
29. Cardozo-Junior F. M., Carneiro R. F. V., Goto B. T., Bezerra A. A. C., Araújo A. S. F., Nunes L. A. P. L. (2012). Arbuscular mycorrhizal fungi in degraded lands in Northeast Brazil. *Afr. J. Microbiol. Res.* 6 7198–7205. 10.5897/AJMR12.1132
30. CBD (2010). *Aichi Biodiversity Targets.* Available at: http://www.cbd.int/sp/targets/

31. Chagnon P.-L., Bradley R. L., Maherali H., Klironomos J. N. (2013). A trait-based framework to understand life history of mycorrhizal fungi. *Trends Plant Sci.* 18 484–491. 10.1016/j.tplants.2013.05.001
32. Choi Y. D. (2004). Theories for ecological restoration in changing environment: toward 'futuristic' restoration. *Ecol. Res.* 19 75–81. 10.1111/j.1440-1703.2003.00594.x
33. Cortines E., Valcarcel R. (2009). Influence of pioneer-species combinations on restoration of disturbed ecosystems in the Atlantic Forest, Rio de Janeiro, Brazil. *Rev. Arvore* 33 927–936. 10.1590/S0100-67622009000500015
34. Cranenbrouck S., Voets L., Bivort C., Renard L., Strullu D.-G., Declerck S. (2005). "Methodologies for in vitro cultivation of arbuscular mycorrhizal fungi with root organs," in *Soil Biology In Vitro Culture of Mycorrhizae* Vol. 4 eds Declerck S., Strullu D. G., Fortin A., editors. (Berlin: Springer-Verlag;), 341–375.
35. Dag A., Yermiyahu U., Ben-Gal A., Zipori I., Kapulnik Y. (2009). Nursery and post-transplant field response of olive trees to arbuscular mycorrhizal fungi in an arid region. *Crop Pasture Sci.* 60 427–433. 10.1071/CP08143
36. Davison J., Öpik M., Daniell T. J., Moora M., Zobel M. (2011). Arbuscular mycorrhizal fungal communities in plant roots are not random assemblages. *FEMS Microb. Ecol.* 78 103–115. 10.1111/j.1574-6941.2011.01103.x
37. Duponnois R., Ouahmane L., Kane A., Thioulouse J., Hafidi M., Boumezzough A., et al.,. (2011). Nurse shrubs increased the early growth of Cupressus seedlings by enhancing belowground mutualism and soil microbial activity. *Soil Biol. Biochem.* 43 2160–2168. 10.1016/j.soilbio.2011.06.020
38. Egler F. E. (1954). Vegetation science concepts. I. Initial floristic composition, a factor in old-field vegetation development. *Vegetatio* 4412–417. 10.1007/BF00275587
39. Figueiredo M. D. B., Seldin L., de Araujo F. F., Mariano R. D. R. (2010). "Plant growth promoting rhizobacteria: fundamentals and applications," in *Plant Growth and Health Promoting Bacteria. Microbiology Monographs*, ed. Maheshwari D. K., editor. (Berlin:

Springer-Verlag;), 21–43.

40. File A. L., Klironomos J. N., Maherali H., Dudley S. A. (2012). Plant kin recognition enhances abundance of symbiotic microbial partner. *PLoS ONE* 7:e45648 10.1371/journal.pone.0045648

41. Finlay R. D. (2008). Ecological aspects of mycorrhizal symbiosis: with special emphasis on the functional diversity of interactions involving the extraradical mycelium. *J. Exp. Bot.* 59 1115–1126. 10.1093/jxb/ern059

42. Fischer C. R., Janos D. P., Perry D. A., Liberman R. G., Sollins P. (1994). Mycorrhiza inoculum potentials in tropical secondary succession. *Biotropica* 26 369–377. 10.2307/2389230

43. Fortin J. A., Declerck S., Strullu D.-G. (2005). "In vitro culture of mycorrhizas," in *Soil Biology In Vitro Culture of Mycorrhizae* Vol. 4eds Declerck S., Strullu D. G., Fortin A., editors. (Berlin: Springer-Verlag;), 3–14.

44. Friese C. F., Allen M. F. (1991). The spread of VA Mycorrhizal fungal hyphae in the soil: inoculum types and external hyphae architecture. *Mycologia* 83 409–418. 10.2307/3760351

45. Garcia K., Zimmermann S. D. (2014). The role of mycorrhizal associations in plant potassium nutrition. *Front. Plant Sci.* 5:33710.3389/fpls.2014.00337

46. Gianinazzi S., Gollotte A., Binet M.-N., van Tuinen D., Redecker D., Wipf D. (2010). Agroecology: the key role of arbuscular mycorrhizas in ecosystem services. *Mycorrhiza* 20 519–530. 10.1007/s00572-010-0333-3

47. Gleason H. A. (1927). Further views on the succession-concept. *Ecology* 8 229–326. 10.2307/1929332

48. Gómez-Aparicio L. (2009). The role of plant interactions in the restoration of degraded ecosystems: a meta-analysis across life-forms and ecosystems. *J. Ecol.* 97 1202–1214. 10.1111/j.1365-2745.2009.01573.x

49. Good M. K., Clarke P. J., Price J. N., Reid N. (2014). Seasonality and facilitation drive tree establishment in a semi-arid floodplain savanna. *Oecologia* 175 261–271. 10.1007/s00442-014-2886-x

50. Gotzenberger L., de Bello F., Brathen K. A., Davison J., Dubuis

A., Guisan A., et al.,. (2011). Ecological assembly rules in plant communities-approaches, patterns and prospects. *Biol. Rev.* 87 111–127. 10.1111/j.1469-185X.2011.00187.x

51. Govindarajulu M., Pfeffer P. E., Jin H., Abubaker J., Douds D. D., Allen J. W., et al.,. (2005). Nitrogen transfer in the arbuscular mycorrhizal symbiosis. *Nat. Lett.* 435 819–823. 10.1038/nature03610
52. Gutjahr C. (2014). Phytohormone signaling in arbuscular mycorrhiza development. *Curr. Opin. Plant Biol.* 20 26–34. 10.1016/j.pbi.2014.04.003
53. Habte M. (1989). Impact of simulated erosion on the abundance and activity of indigenous vesicular-arbuscular mycorrhizal endophytes in an Oxisol. *Biol. Fer. Soils* 7 164–167. 10.1007/BF00292576
54. Habte M., Miyasaka S. C., Matsuyama D. T. (2001). "Arbuscular mycorrhiza fungi improve early forest-tree establishment," in *Plant Nutrition: Food Security and Sustainability of Agroecosystems Through Basic and Applied Research*, eds Horst W. J., Schenk M. K., Bürkert A., Claassen N., Flessa H., Frommer W. B., et al., editors. (Dordrecht: Kluwer Academic Publishers;), 644–645.
55. Hailemariam M., Birhane E., Asfaw Z., Zewdie S. (2013). Arbuscular mycorrhizal association of indigenous agroforestry tree species and their infective potential with maize in the rift valley, Ethiopia. *Agrofor. Sys.* 87 1–14. 10.1007/s10457-013-9634-9
56. Harris J. A., Hobbs R. J., Higgs E., Aronson J. (2006). Ecological restoration and global climate change. *Restor. Ecol.* 14 170–176. 10.1111/j.1526-100X.2006.00136.x
57. Hartnett D. C., Wilson G. W. T. (1999). Mycorrhiza influence plant community structure and diversity in tall grass Prairie. *Ecology* 801187–1195. 10.1890/0012-9658(1999)080[1187:MIPCSA]2.0.CO;2
58. Heneghan L., Miller S. P., Baer S., Callaham M. A., Montgomery J., Pavao-Zuckerman M., et al.,. (2008). Integrating soil ecological knowledge into restoration management. *Restor. Ecol.* 16 608–617. 10.1111/j.1526-100X.2008.00477.x
59. Herman D. J., Firestone M. K., Nuccio E., Hodge A. (2012). Interactions between an arbuscular mycorrhizal fungus and

a soil microbial community mediating litter decomposition. *FEMS Microbiol. Ecol.* 80 236–247. 10.1111/j.1574-6941.2011.01292.x

60. Higgs E., Falk D. A., Guerrini A., Hall M., Harris J., Hobbs R. H., et al.,. (2014). The changing role of history in restoration ecology. *Front. Ecol. Environ.* 12:499–506. 10.1890/110267
61. Hildebrandt U., Regvar M., Bothe H. (2007). Arbuscular mycorrhiza and heavy metal tolerance. *Phytochemistry* 68 139–146. 10.1016/j.phytochem.2006.09.023
62. Hobbs R. J., Jentsch A., Temperton V. M. (2007). "Restoration as a process of assembly and succession mediated by disturbance," in *Linking Restoration and Ecological Succession*, eds Walker L. R., Walker J., Hobbs R. J., editors. (New York, NY: Springer;), 150–167.
63. Hoeksema J. D., Chaudhary V. B., Gehring C. A., Johnson N. C., Karst J., Koide R. T., et al.,. (2010). A meta-analysis of context-dependency in plant response to inoculation with mycorrhizal fungi. *Ecol. Lett.* 13 267–407. 10.1111/j.1461-0248.2009.01430.x
64. Holl K. D., Zahawi R. A., Cole R. J., Ostertag R., Cordell S. (2010). Planting seedlings in tree islands versus plantations as a large-scale tropical forest restoration strategy. *Restor. Ecol.* 19 470–479. 10.1111/j.1526-100X.2010.00674.x
65. Hrynkiewicz K., Baum C. (2011). "The potential of rhizosphere microorganisms to promote the plant growth in disturbed soils," in *Environmental Protection Strategies for Sustainable Development. Strategies for Sustainability*, eds Malik A., Grohmann E., editors. (New York, NY: Springer Science+Business;), 35–64.
66. Huante P., Ceccon E., Orozco-Segovia A., Sánchez-Coronado M. E., Acosta I., Rincón E. (2012). The role of arbuscular mycorrhizal fungi on the early stage restoration of seasonally dry tropical rain forest in Chamela, Mexico. *Rev. Árvore Viçosa-MG* 36 279–289. 10.1590/S0100-67622012000200009
67. Ijdo M., Cranenbrouck S., Declerck S. (2011). Methods for large-scale production of AM fungi: past, present, and future. *Mycorrhiza* 211–16. 10.1007/s00572-010-0337-z
68. Jacobs D. F., Oliet J. A., Aronson J., Bolte A., Bullock J. M., Donoso P. J., et al.,. (2015). Restoring forests: what constitutes success in

the twenty-first century? *New For.* 46 601–614. 10.1007/s11056-015-9513-5

69. Janos D. P. (1980). Mycorrhizae influence tropical succession. *Biotropica* 12 56–64. 10.1007/s00572-012-0464-9
70. Janos D. P., Scott J., Aristiza'bal C., Bowman D. M. J. S. (2013). Arbuscular-mycorrhizal networks inhibit *Eucalyptus tetrodonta*seedlings in rain forest soil microcosms. *PLoS ONE* 8:e57716 10.1371/journal.pone.0057716
71. Janoušková M., Krak K., Wagg C., Štorchová H., Caklová P., Vosátkaa M. (2013). Effects of inoculum additions in the presence of a preestablished arbuscular mycorrhizal fungal community. *Appl. Environ. Microbiol.* 79 6507–6515. 10.1128/AEM.02135-13
72. Jasper D. A., Abbott L. K., Robson A. D. (1989). Hyphae of a vesicular-arbuscular mycorrhizal fungus maintain infectivity in dry soil, except when the soil is disturbed. *New Phytol.* 112 101–107. 10.1111/j.1469-8137.1989.tb00314.x
73. Jeffries P., Craven-Griffiths A., Barea J. M., Levy Y., Dodd J. C. (2002). "Application of arbuscular mycorrhizal fungi in the revegetation of desertified Mediterranean ecosystems," in *Mycorrhizal Technology in Agriculture*, eds Gianinazzi S., Schuepp H., Barea J. M., Haselwandter K., editors. (Basel: Birkhauser verlag;), 151–174.
74. Johnson N. C., Graham J. H., Smith F. A. (1997). Functioning of mycorrhizal associations along the mutualism parasitism continuum. *New Phytol.* 135 575–586. 10.1046/j.1469-8137.1997.00729.x
75. Kapulnik Y., Tsror L., Zipori I., Hazanovsky M., Wininger S., Dag A. (2010). Effect of AMF application on growth, productivity and susceptibility to Verticillium wilt of olives grown under desert conditions. *Symbiosis* 52 103–111. 10.1007/s13199-010-0085-z
76. Karthikeyan A., Krishnakumar N. (2012). Reforestation of bauxite mine spoils with Eucalyptus tereticornis Sm. seedlings inoculated with arbuscular mycorrhizal fungi. *Ann. For. Res.* 55 207–216.
77. Kiers E. T., Lovelock C. E., Krueger E. L., Herre E. A. (2000). Differential effects of tropical arbuscular mycorrhizal fungal inocula on root colonization and tree seedling growth: implications for tropical forest diversity. *Ecol. Lett.* 3 106–113. 10.1046/j.1461-0248.2000.00126.x

78. Kikvidze Z., Armas C., Fukuda K., Martínez-García L. B., Miyata M., Oda-Tanaka A., et al.,. (2010). The role of arbuscular mycorrhizae in primary succession: differences and similarities across habitats. *Web Ecol.* 10 50–57. 10.5194/we-10-50-2010

79. Kikvidze Z., Brooker R. W., Butterfield B. J., Callaway R. M., Cavieres L. A., Cook B. J., et al.,. (2015). The effects of foundation species on community assembly: a global study on alpine cushion plant communities. *Ecology* 96 2064–2069. 10.1890/14-2443.1

80. Klironomos J. N. (2003). Variation in plant response to native and exotic mycorrhizal fungi. *Ecology* 84 2292–2301. 10.1890/02-0413

81. Klironomos J. N., McCune J., Hart M., Neville J. (2000). The influence of arbuscular mycorrhiza on the relationship between plant diversity and productivity. *Ecol. Lett.* 3 137–141. 10.1046/j.1461-0248.2000.00131.x

82. Koide R. T., Mosse B. (2004). A history of research on arbuscular mycorrhiza. *Mycorrhiza* 14 145–163. 10.1007/s00572-004-0307-4

83. Kula A. A. R., Hartnett D. C., Wilson G. W. T. (2005). Effects of mycorrhizal symbiosis on tall grass prairie plant–herbivore interactions.*Ecol. Lett.* 8 61–69. 10.1111/j.1461-0248.2004.00690.x

84. Kumar A., Raghuwanshi R., Upadhyay R. S. (2010). Arbuscular mycorrhizal technology in reclamation and revegetation of coal mine spoils under various revegetation models. *Engineering* 2 683–689. 10.4236/eng.2010.29088

85. Lamb D., Erskine P. D., Parrotta J. D. (2005). Restoration of degraded tropical forest landscapes. *Science* 310 1628–1632. 10.1126/science.1111773

86. Larcher W. (1995). *Physiological Plant Ecology.* Berlin: Springer-Verlag.

87. LeeE.-H.,EoJ.-K.,KaK.-H.,EomA.-H.(2013).Diversityofarbuscular mycorrhizal fungi and their roles in ecosystems. *Mycobiology*41 121–125. 10.5941/MYCO.2013.41.3.121

88. Leifheit E. F., Verbruggen E., Rillig M. C. (2014). Arbuscular mycorrhizal fungi reduce decomposition of woody plant litter while increasing soil aggregation. *Soil Biol. Biochem.* 81 323–328. 10.1016/j.soilbio.2014.12.003

89. Lekberg Y., Koide R. T. (2005). Is plant performance limited by abundance of arbuscular mycorrhizal fungi? A meta-analysis of studies published between 1988 and 2003. *New Phytol.* 168 189–204. 10.1111/j.1469-8137.2005.01490.x
90. Lerat S., Lapointe L., Gutjahr S., Piche Y., Vierheilig H. (2003). Carbon partitioning in a split-root system of arbuscular mycorrhizal plants is fungal and plant species dependent. *New Phytol.* 157 589–595. 10.1046/j.1469-8137.2003.00691.x
91. Leyval C., Turnau K., Haselwandter K. (1997). Effect of heavy metal pollution on mycorrhizal colonization and function: physiological, ecological and applied aspects. *Mycorrhiza* 7 139–153. 10.1007/s005720050174
92. Lin G., McCormack M. L., Guo D. (2015). Arbuscular mycorrhizal fungal effects on plant competition and community structure. *J. Ecol.*103 1224–1232. 10.1111/1365-2745.12429
93. Mack R. N., Simberloff D., Lonsdale W. M., Evans H., Clout M., Bazzaz F. A. (2000). Biotic invasions: causes, epidemiology, global consequences and control. *Ecol. Appl.* 10 689–710. 10.2307/2641039
94. Maherali H., Klironomos J. N. (2007). Influence of phylogeny on fungal community assembly and ecosystem functioning. *Science* 316 1746–1748. 10.1126/science.1143082
95. Manaut N., Sanguin H., Ouahmane L., Bressan M., Thioulouse J., Baudoin E., et al.,. (2015). Potentialities of ecological engineering strategy based on native arbuscular mycorrhizal community for improving afforestation programs with carob trees in degraded environments. *Ecol. Eng.* 79 113–119. 10.1016/j.ecoleng.2015.03.007
96. Marschner H. (1995). *Mineral Nutrition of Higher Plants*, 2nd Edn London: Academic Press.
97. Michelsen A. (1992). Mycorrhiza and root nodulation in tree seedlings from five nurseries in Ethiopia and Somalia. *For. Ecol. Manag.* 48335–344. 10.1016/0378-1127(92)90154-2
98. Miyasaka S. C., Habte M., Friday J. B., Johnson E. V. (2003). *Manual on Arbuscular Mycorrhizal Fungus Production and Inoculation Techniques.* Honolulu: CTAHR, University of Hawaii.
99. Navarro J. M., Perez-Tornero O., Morte A. (2013). Alleviation

of salt stress in citrus seedlings inoculated with arbuscular mycorrhizal fungi depends on the rootstock salt tolerance. *J. Plant Physiol.* 171 76–85. 10.1016/j.jplph.2013.06.006

100. Odum E. P. (1969). The strategy of ecosystem development. *Science* 164 262–270. 10.1126/science.164.3877.262
101. Onguene N. A., Kuyper T. W. (2005). Growth response of three native timber species to soils with different arbuscular mycorrhizal inoculums potentials in South Cameroon Indigenous inoculum and effect of addition of grass inoculums. *For. Ecol. Manag.* 210 283–290. 10.1016/j.foreco.2005.02.038
102. Osoria N. W., Habte M. (2001). Synergetic influence of an arbuscular mycorrhizal fungus and P solubilizing fungus on growth and P uptake of Leucaena leucocephala in an oxisol. *Arid Land Res. Manag.* 1115 263–274. 10.1080/15324980152119810
103. Ouahmane L., Hafidi M., Thioulouse J., Ducousso M., Kisa M., Prin Y., et al.,. (2006). Improvement of *Cupressus atlantica* Gaussen growth by inoculation with native arbuscular mycorrhizal fungi. *J. Appl. Microbiol.* 103 683–690. 10.1111/j.1365-2672.2007.03296.x
104. Padilla F. M., Ortega R., Sánchez J., Pugnaire F. I. (2009). Rethinking species selection for restoration of arid shrublands. *Basic Appl. Ecol.* 10 640–647. 10.1016/j.baae.2009.03.003
105. Palmer M. A., Ambrose R. F., Poff N. L. (1997). Ecological theory and community restoration ecology. *Restor. Ecol.* 5 291–300. 10.1046/j.1526-100X.1997.00543.x
106. Parniske M. (2008). Arbuscular mycorrhiza: the mother of plant root endosymbioses. *Microbiology* 6 763–775. 10.1038/nrmicro1987
107. Perry D. A., Molina R., Amaranthus M. P. (1987). Mycorrhizae, mycorhizosphere, and reforestation: current knowledge and research need. *Can. J. For. Res.* 17 929–940.
108. Pouyu-Rojas E., Siqueira J. O. (2000). Arbuscular mycorrhizal and soil fertilization on post-transplant development of out plants of seven forest species. *Pesqui. Agropecu. Bras.* 35 103–114. 10.1590/S0100-204X2000000100013
109. Puri A., Adholeya A. (2013). A new system using Solanum tuberosum for the co-cultivation of *Glomus intraradices* and its potential for mass producing spores of arbuscular mycorrhizal

fungi. *Symbiosis* 59 87–97. 10.1007/s13199-012-0213-z

110. Quoreshi A. M. (2008). "The use of mycorrhizal biotechnology in restoration of disturbed ecosystem," in *Mycorrhizae: Sustainable Agriculture and Forestry*, eds Siddiqui Z. A., Akhatar M. S., Futai K., editors. (New York, NY: Springer Science + Business Media B.V;), 303–320.

111. Redecker D., Raab P. A. (2006). Phylogeny of the Glomeromycota (arbuscular mycorrhizal fungi): recent developments and new gene markers. *Mycologia* 98 885–895. 10.3852/mycologia.98.6.885

112. Renker C., Zobel M., Öpik M., Allen M. F., Allen E. B., Vosátka M., et al.,. (2004). "Structure, dynamics, and restoration of plant communities: do arbuscular mycorrhizae matter?," in *Assembly Rules and Restoration Ecology : Bridging the Gap Between Theory and Practice*, eds Temperton V. M., Hobbs R. J., Nuttle T., Halle S., editors. (Washington, DC: Island Press;), 189–229.

113. Requena N., Perez-Solis E., Azcón-Aguilar C., Jeffries P., Barea J. M. (2001). Management of indigenous plant-microbe symbioses aids restoration of desertified ecosystems. *Appl. Environ. Microbiol.* 67 495–498. 10.1128/AEM.67.2.495-498.2001

114. Riginos C. (2009). Grass competition suppresses savanna tree growth across multiple demographic stages. *Ecology* 90 335–340. 10.1890/08-0462.1

115. Rilling M. C. (2004). Arbuscular mycorrhizae and terrestrial ecosystem processes. *Ecol. Lett.* 7 740–754. 10.1111/j.1461-0248.2004.00620.x

116. Rodriguez-Echeverria S., Costa S. R., Freitas H. (2007). "Biodiversity and interactions in the rhizosphere: effects on ecosystem functioning," in *Functional Plant Ecology*, 2nd Edn, eds Pugnaire F. I., Valladares F., editors. (New York: CRC Press;), 581–600.

117. Ruiz-Jaen M. C., Aide T. M. (2005). Restoration success: how is it being measured? *Restor. Ecol.* 13 569–577. 10.1111/j.1526-100X.2005.00072.x

118. Saharan B. S., Nehra V. (2011). Plant growth promoting rhizobacteria: a critical review. *Life Sci. Med. Res.* 21 1–30.

119. Sanon A., Ndoye F., Baudoin E., Prin Y., Galiana A., Duponnois R.

(2010). "Management of the mycorrhizal soil infectivity to improve reforestation programs' achievements in Sahelian ecosystems," in *Current Research, Technology and Education Topics in Applied Microbiology and Microbial Biotechnology* Vol. 2 ed. Méndez-Vilas A., editor. (Badajoz: FORMATEX Research Center;), 230–238.

120. Scheublin T. R., van Logtestijn R. S. P., van der Heijden M. G. A. (2007). Presence and identity of arbuscular mycorrhizal fungi influence competitive interactions between plant species. *J. Ecol.* 95 631–638. 10.1111/j.1365-2745.2007.01244.x
121. Schnitzer S. A., Klironomos J. N., Lambers J. H. R., Kinkel L. L., Reich P. B., Xiao K., et al.,. (2011). Soil microbes drive the classic plant diversity–productivity pattern. *Ecology* 92 296–303. 10.1890/10-0773.1
122. Schnoor T. K., Lekberg Y., Rosendahl S., Olsson P. A. (2011). Mechanical soil disturbance as a determinant of arbuscular mycorrhizal fungal communities in semi-natural grassland. *Mycorrhiza* 21 211–220. 10.1007/s00572-010-0325-3
123. Schüßler A., Martin H., Cohen D., Fitz M., Wipf D. (2007). Arbuscular mycorrhiza: studies on the *Geosiphon* symbiosis lead to the characterization of the first Glomeromycotan sugar transporter. *Plant Signal. Behav.* 2 431–434. 10.4161/psb.2.5.4465
124. SER (2004). *The SER International Primer on Ecological Restoration. Society for Ecological Restoration International Science & Policy Working Group Version 2 Number 1.* Available online: http://www.ser.org/resources/resources-detail-view/ser-international-primer-on-ecological-restoration
125. Simard S. W., Austin M. E. (2010). "The role of mycorrhizas in forest soil stability with climate change," in *Climate Change and Variability*, ed. Simard S. W., editor. (Rijeka: InTech Open Access;), 275–302.
126. Skujins J., Allen M. F. (1986). Use of mycorhiza for land rehabilitation. *MIRCEN J. Microbiol. Biotechnol.* 2 161–176. 10.1007/BF00937191
127. Smith S. E., Jakobsen I., Grønlund M., Smith F. A. (2011). Roles of arbuscular mycorrhizas in plant phosphorus nutrition: interactions between pathways of phosphorus uptake in arbuscular mycorrhizal

roots have important implications for understanding and manipulating plant phosphorus acquisition. *Plant Physiol.* 156 1050–1057. 10.1104/pp.111.174581

128. Smith S. E., Smith F. A. (2011). Roles of arbuscular mycorrhizas in plant nutrition and growth: new paradigms from cellular to ecosystems scales. *Annu. Rev. Plant Biol.* 63 227–250. 10.1146/annurev-arplant-042110-103846
129. Soares C. R. F. S., Siqueira J. O. (2008). Mycorrhiza and phosphate protection of tropical grass species against heavy metal toxicity in multi-contaminated soil. *Biol. Fert. Soils* 44 833–841. 10.1007/s00374-007-0265-z
130. Soka G., Ritchie M. (2014). Arbuscular mycorrhizal symbiosis and ecosystem processes: prospects for future research in tropical soils. *Open J. Ecol.* 4 11–22. 10.4236/oje.2014.41002
131. Song Y. Y., Ye M., Li C., He X., Zhu-Salzman K., Wang R. L., et al.,. (2014). Hijacking common mycorrhizal networks for herbivore-induced defence signal transfer between tomato plants. *Sci. Rep.* 4 1–8. 10.1038/srep03915
132. Taylor T. N., Remy W., Hass H., Kerp H. (1995). Fossil arbuscular mycorrhizae from the early Devonian. *Mycologia* 87 560–573. 10.2307/3760776
133. Thomas E., Jalonen R., Loo J., Boshier D., Gallo L., Cavers S., et al.,. (2014). Genetic considerations in ecosystem restoration using native tree Species. *For. Ecol. Manag.* 333 66–75. 10.1016/j.foreco.2014.07.015
134. Timonen S., Marschner P. (2006). "Mycorrhizosphere concept," in *Soil Biology: Microbial Activity in the Rhizosphere* Vol. 7 eds Mukerji K. G., Manoharachary C., Singh J., editors. (Berlin: Springer-Verlag;), 155–172.
135. UN (2015). *Transforming Our World: The 2030 Agenda for Sustainable Development.* Available at: https://sustainabledevelopment.un.org/post2015/transformingourworld
136. Van der Heijden M. G. A., Klironomos J. N., Ursic M., Moutoglis P., Streitwolf-Engel R., Boller T., et al.,. (1998). Mycorrhizal fungal diversity determines plant biodiversity, ecosystem variability and productivity. *Nature* 396 69–72. 10.1038/23932

137. Van der Heijden M. G. A., Klironomos J. N., Ursic M., Moutoglis P., Streitwolf-Engel R., Boller T., et al.,. (1999). "Sampling effect," a problem in biodiversity manipulation? A reply to David A. Wardle. *Oikos* 87 408–410. 10.2307/3546758

138. Veiga R. S. L., Jansa J., Frossard E., van der Heijden M. G. A. (2011). Can arbuscular mycorrhizal fungi reduce the growth of agricultural weeds? *PLoS ONE* 6:e27825 10.1371/journal.pone.0027825

139. Vogelsang K. M., Reynolds H. L., Bever J. D. (2006). Mycorrhizal fungal identity and richness determine the diversity and productivity of a tall grass prairie system. *New Phytol.* 172 554–562. 10.1111/j.1469-8137.2006.01854.x

140. Walder F., Niemann H., Natarajan M., Lehmann M. F., Boller T., Wiemken A. (2012). Mycorrhizal networks: common goods of plants shared under unequal terms of trade. *Plant Physiol.* 159 789–797. 10.1104/pp.112.195727

141. Wang B., Qiu Y. L. (2006). Phylogenetic distribution and evolution of mycorrhizas in land plants. *Mycorrhiza* 16 299–363. 10.1007/s00572-005-0033-6

142. Weremijewicz J., Janos D. P. (2013). Common mycorrhizal networks amplify size inequality in *Andropogon gerardii* monocultures. *New Phytol.* 198 203–213. 10.1111/nph.12125

143. World Resource Institute (2015). *What is Degraded Land?* Available at: http://www.wri.org/faq/what-degraded-land

144. Wubet T., Kottke I., Teketay D., Oberwinkler F. (2009). Arbuscular mycorrhizal fungal community structures differ between co-occurring tree species of dry Afromontane tropical forest, and their seedlings exhibit potential to trap isolates suited for reforestation. *Mycol. Prog.* 8 317–328. 10.1007/s11557-009-0602-8

145. Wubet T., Weiß M., Kottke I., Oberwinkler F. (2006). Two threatened coexisting indigenous conifer species in the dry afromontane forests of Ethiopia are associated with distinct arbuscular mycorrhizal communities. *Can. J. Bot.* 84 1617–1627. 10.1139/b06-121

146. Yang H., Dai Y., Wang X., Zhang Q., Zhu L., Bian X. (2014). Meta-analysis of interactions between arbuscular mycorrhizal fungi and biotic stressors of plants. *Sci. World J.* 2014 1–7.

10.1155/2014/746506

147. Young T. P., Chase J. M., Huddleston R. T. (2001). Community succession and assembly: comparing, contrasting and combining paradigms in the context of ecological restoration. *Ecol. Restor.* 19 5–18.

148. Young T. P., Petersen D. A., Clary J. J. (2005). The ecology of restoration: historical links, emerging issues and unexplored realms. *Ecol. Lett.* 8 662–673. 10.1111/j.1461-0248.2005.00764.x

149. Zedler J. B. (2005). Ecological restoration: guidance from theory. *San Francisco Estuar. Watershed Sci.* 3 1–31. 10.15447/sfews.2005v3iss2art4

150. Zobel M., pik M. (2014). Plant and arbuscular mycorrhizal fungal (AMF) communities – which drives which? *J. Veg. Sci.* 25 1133–1140. 10.1111/jvs.12191

11
CHAPTER

EFFECT OF 40 AND 80 YEARS OF CONIFER REGROWTH ON SOIL MICROBIAL ACTIVITIES AND COMMUNITY STRUCTURE IN SUBTROPICAL LOW MOUNTAIN FORESTS

Ed-Haun Chang [1], Tsai-Huei Chen [2], Guang-long Tian [3], Chun-Kai Hsu [2] and Chih-Yu Chiu [4]

[1]Mackay Junior College of Medicine, Nursing and Management, Beitou, Taipei 112, Taiwan

[2]Taiwan Forestry Research Institute, Taipei 10066, Taiwan

[3]Environmental Monitoring and Research Division, Monitoring and Research Department, Metropolitan Water Reclamation District of Greater Chicago, IL 60804, USA

[4]Biodiversity Research Center, Academia Sinica, Nangang, Taipei 11529, Taiwan

ABSTRACT

The effects of long-term reforestation on soil microbial communities and biomass are poorly understood. This study was conducted on two coniferous plantations: *Cunninghamia konishii* Hayata, planted 40 years ago (CONIF-40), and *Calocedrus formosana* (Florin) Florin, planted 80 years ago (CONIF-80). An adjacent natural broadleaf forest (BROAD-Nat) was used as a control. We determined microbial biomass C and N contents, enzyme activities, and community composition (via phospholipid fatty acid [PLFA] assessment). Both microbial biomass and PLFA content were higher in the summer than in the winter and differed among the forests in summer only. Total PLFA, total bacterial, gram-positive bacterial, gram-negative bacterial, and vesicular arbuscular mycorrhizal fungal contents followed the same pattern. Total fungal content and the ratios of fungi to bacteria and of gram-positive to gram-negative bacteria were highest in CONIF-40, with no difference between the other forests. Principal component analysis of PLFA contents revealed that CONIF-40 communities were distinct from those of CONIF-80 and BROAD-Nat. Our results suggest that vegetation replacement during reforestation exerts a prolonged impact on the soil microbial community. The understory broadleaf shrubs and trees established after coniferous plantation reforestation may balance out the effects of coniferous litter, contributing to bacterial recovery.

Keywords:

PLFA; microbial community; soil enzymes; forest

INTRODUCTION

The management of forest ecosystems, such as harvesting, burning, deforestation, and reforestation, results not only in quantitative and qualitative changes in soil C and N contents but also in eventual changes in the microbial community [1,2]. Tree type is a primary factor affecting the microbial community structure in forest soils as it controls the input of substrate type to soil [3] and soil chemical properties, such as the

C content, C/N ratio, and pH [4]. Reforestation with exotic and mostly single species at original natural forest sites has changed the soil microbial community in subtropical areas [5,6]. Ushio et al. [7] indicated that tree species had a direct impact on the soil microbial community in tropical montane forests. Likewise, these authors suggested that coniferous forest soils exhibited higher ratios of gram-positive (G+) to gram-negative (G–) bacteria and fungi to bacteria than did broadleaf forests.

After reforestation, the microbial community structure evolves over time to approach that in native forest soils. Using phospholipid fatty acids (PLFA) analysis, Moore-Kucera and Dick [8] found that soil total PLFA, as well as total bacteria and fungi were lower in an 8-year-old replanted forest than in old-growth forests. However, those in a 25-year-old replanted site were similar to those of an old-growth forest. Considerable research has been conducted to assess the response of the soil microbial community under secondary forests in temperate regions [9]. PLFA has been used to estimate changes in the soil microbial community after forest management [10]. Comparatively less attention has been paid to this subject in subtropical forests, where higher temperatures and precipitation may elicit a different response from the soil microbial community under secondary forests [11,12]. Furthermore, little information is available on the extent to which microbial communities differ between native forests and 50 to 100-year-old reforested areas.

Many studies have reported that coniferous and broadleaf tree species exert strong influences on soil microbial communities [7,13]. However, a recent study by Wang et al. [14] showed that differences in bacterial community structure between coniferous and broadleaf forests disappeared due to changes in soil properties. In an attempt to promote the economical production of timber, reforested conifers replaced several native broadleaf hardwood forests decades ago in a low- to high-elevation subtropical montane region in central Taiwan. Our previous study suggested that disturbances during forest conversion influence the soil bacterial community, eliciting differences among three forest ecosystems in the same climate [15]. The results of Lin et al. [15] also revealed that soil bacterial community diversity in 40-year-old coniferous forest was greater than that in 80-year-old coniferous forest, which was in turn greater than that in natural broadleaf forest, reflecting a trend of disturbance.

We hypothesized that the soil microbial community of secondary forests would establish a similar structure to that of a natural system of a different vegetation type after a sufficient period of time. We therefore compared the soil microbial community structure in a 40-year-old and an 80-year-old secondary coniferous forest with that of a native broadleaf forest. We used PLFA analysis to provide quantitative data on the structure of microbial communities [16,17] and evaluated similarities in community structure across the different forest types. We found that while soil microbial communities differed between native broadleaf forests and 40-year-old coniferous forests, such differences were reduced between native broadleaf and 80-year-old coniferous forests. The understory broadleaf shrubs and trees established after the long-term reforestation of coniferous plantations may contribute to the recovery of bacteria in these secondary forests.

MATERIALS AND METHODS

Site Description

This study was conducted at the Lienhuachi Experimental Forest in a subtropical low mountain area in central Taiwan (23°54′ N, 120°54′ E). The elevation ranges from 600 to 950 m above sea level, with a mean annual precipitation of 2200 mm and a mean annual temperature of 20.8 °C. Soils are moderately well drained and classified as Typic Dystrudept (U.S. Soil Taxonomy). As one of the few remaining low-elevation, natural, undisturbed forests in central Taiwan, the site is part of a long-term ecological study covering 461 ha of a low-altitude forest ecosystem, including 261 ha belonging to evergreen natural broadleaf hardwood forest dominated by Fagaceae and Lauraceae species. Coniferous plantations established 80 and 40 years ago with *Calocedrus formosana* (Florin) Florin and *Cunninghamia konishii* Hayata, respectively, replaced approximately 50% of the native forests on this study site.

The study included the two coniferous plantations, i.e. the 40-year-old *C. konishii* plantation (CONIF-40) and the 80-year-old *C. formosana* plantation (CONIF-80), as well as a natural broadleaf forest (BROAD-Nat) as the control. The CONIF-80 and CONIF-40 plantations, which were established by reforestation with coniferous trees after the native forest was cleared, encompass 5 and 100 ha, respectively. The BROAD-Nat forest exhibits a high diversity of plant species and an even species composition dominated by tree species in the families Lauraceae and Fagaceae. BROAD-Nat trees exhibit a mean height of 10–18 m, a density of 1450 stems/ha, and a diameter at breast height (DBH) of 10–30 cm (Table 1). Several hardwood tree species in the BROAD-Nat forest form a canopy structure with multiple layers.

Table 1: Indices of natural broadleaf forest (BROAD-Nat), 80-year-old *Calocedrus*(CONIF-80) plantation, and 40-year-old *Cunninghamia* (CONIF-40) plantation at Lienhuachi Experimental Forest

Forest type	**Species**	**Density (stems·ha^{-1})**	**Plant Height (m)**	**Average DBH ^{1}(cm)**	**Average Basal Area (m^{2}·ha^{-1})**	**Wood Volume (m^{3}·ha^{-1})**
BROAD-Nat	*Cryptocarya chinensis* (Hance) Hemsl.	250	10.9	15.1	7.89	77.28
	Litsea acuminatae (Blume) Kurata	175	12.5	14.7	4.74	41.85
	Prunus phaeosticta (Hance) Maxim.	175	13.2	16.7	4.64	36.20
	Lithocarpus amygdalifolius (Skan) Hayata	100	17.5	18.5	3.84	33.20
	Schefflera octophylla (Lour.) Harms	100	13.5	16.8	2.56	16.45
	Cinnamomum subavenium Miq.	175	13.1	10.5	2.44	18.44
	Cyclobalanopsis pachyloma (O. Seem.) Schott.	75	10.7	20.3	2.44	19.17
	Styrax suberifolia Hook. & Arn.	75	16.3	17.2	1.98	16.44
	Pasania harlandii (Hance) Oerst.	25	16.0	31.1	1.90	13.75
	Machilus zuihoensis Hayata	50	18.0	20.4	1.85	16.20
	Others	250	9.5	9.1	1.92	9.35
	Total	1450			36.20	298.3

CONIF-80	*Calocedrus formosana* (Florin) Florin	275	26.6	48.0	51.44	644.9
	Randia cochinchinensis (Lour.) Merr	625	8.6	6.2	1.95	7.80
	Schefflera octophylla (Lour.) Harms	575	8.7	6.2	1.79	7.24
	Prunus phaeosticta (Hance) Maxim.	275	10.2	7.5	1.26	5.96
	Others	375	8.0	5.9	1.15	4.39
	Total	2125			57.59	670.3
CONIF-40	*Cunninghamia lanceolata* (Lamb.) Hook.	900	15.8	21.4	36.01	289.6
	Styrax formosana Matsum	80	10.6	10.7	0.79	4.14
	Schefflera octophylla (Lour.) Harms	40	12.5	15.1	0.75	4.53
	Others	140	12.0	10.6	1.62	8.75
	Total	1160			39.17	307.0

[1] DBH: diameter at breast height.

In the CONIF-80 forest, *C. formosana* is the major species, with a mean height of 27 m, a density of 275 stems/ha, and a DBH of 48 cm. With a space of 6 m × 6 m, the forest has an open canopy allowing sunshine to pass through. Regrowth of broadleaf trees, including *Randia cochinchinensis* (Lour.) Merr., *Schefflera octophylla* (Lour.) Harm, and *Prunus phaeosticta*(Hance) *Maxim*., forms the second layer of the canopy, with a mean height of 10 m and a density of 1850 stems/ha. Although the density of broadleaf regrowth trees in CONIF-80 is quite high, broadleaf trees make up much less of the canopy than do conifers.

In the CONIF-40 plantation, *C. lanceolata* grows to a mean height of 15.8 m, a density of 900 stems/ha, and a mean stand basal area of 36.0 m^2/ha. This plantation also consists of regrowth of broadleaf trees, such as *Styrax formosana*Matsum. and *S. octophylla,* to a height of 10–12 m and a density of 260 stems/ha. The density of broadleaf tree regrowth in the CONIF-40 plantation is lower than that in the CONIF-80 plantation and therefore, there is not a multiple-canopy structure in the CONIF-40 plantation as there is in the CONIF-80 plantation and BROAD-Nat (Figure 1).

(a) (b) (c)

Figure 1: Photographs of (a) natural broadleaf (BROAD-Nat), (b) 80-year-old coniferous (CONIF-80), and (c) 40-year-old coniferous (CONIF-40) forests at the study site.

Soil Sampling and Analysis

Five plots (50 m × 50 m each) were established in each forest site for collection of soil samples. Three soil core samples were collected in each plot to create a composite sample. The soil samples were collected in February 2011 during the winter (dry) season and in August 2011 during the summer (wet) season at a depth of 0–10 cm using a soil auger that was 8 cm in diameter. Roots and litter were manually removed before samples were sieved through a 2-mm sieve. The samples were maintained at 4 °C in the dark. Several subsamples were air-dried and ground in order to analyze their physicochemical properties, such as total organic C and total N. Subsamples of fresh soil were also freeze-dried at −20 °C immediately after field sampling for PLFA analysis. Additional subsamples were weighed and oven-dried at 105 °C to determine moisture content in order to provide an oven-dried basis for comparison of analytical results. Soil organic C (C_{org}) and total N (N_{tot}) contents were determined using a Fisons NA1500 elemental analyzer (ThermoQuest Italia, Milan, Italy). Soil pH values of air-dried soil were measured using a combination glass electrode (soil:water ratio 1:2.5; [18]).

Biochemical Assays

The microbial biomass of each soil sample was determined using the chloroform fumigation extraction method [19]. Organic C retrieved through K_2SO_4 extraction was measured using a total organic carbon analyzer (Model 1010, OI Analytical, College Station, TX, USA) and converted to microbial biomass C (C_{mic}) assuming a conversion factor of 2.22 [20]. The microbial biomass N (N_{mic}) content was calculated from the ninhydrin-reactive N released from the biomass and determined colorimetrically at 560 nm [21].

Urease (EC 3.5.1.5) activity was determined as described in Kandeler and Gerber [22]. Phosphatase (EC 3.1.3.2) activity was determined using the method of Tabatabai and Bremner [23]. Cellulase (EC 3.2.1.4) and xylanase (EC 3.2.1.8) activities were determined using the method of Schinner and von Mersi [24].

PLFA extraction and analysis was performed using the methods described by Frostegard et al. [25]. Lipids were extracted using a single-phase mixture of the chloroform-methanol-citrate system. Phospholipids were then divided into neutral, glycol-, and phospholipids using a solid-phase extraction column and eluting with chloroform, acetone, and methanol, respectively. Next, phospholipids underwent methylation to form fatty acid methyl esters (FAMEs). The FAMEs were analyzed using capillary gas chromatography (GC) and flame ionization detection with the Thermo Fisher TRACE chromatography system (Thermo Fisher Scientific, Waltham, MA, USA). GC was performed according to the method of Chang et al. [26]. The fatty acid nomenclature was established according to Frostegard et al. [25]. The position of double bond is defined by symbol ω from the methyl end of the molecule. The prefix c refers to cis configuration. Iso and anteiso branching are designated by the prefix i or a, respectively. The prefix cy refers to cyclopropane fatty acids. The fatty acids i15:0, a15:0, 15:0, i16:0, 16:1ω7c, 17:0, i17:0, cy17:0, 18:1ω7c, and cy19:0 were categorized as being derived from bacteria: i15:0, a15:0, i16:0, and i17:0 from G+ bacteria; 16:1ω7c, cy17:0, 18:1ω7c, and cy19:0 from G− bacteria; and 10Me18:0 from actinomycetes. The fatty acid 18:2ω6c was categorized as being derived from fungi, and 16:1ω5c was categorized as being derived from vesicular-arbuscular mycorrhizal (VAM) fungi [16,27].

Statistical Analysis

Two-way analysis of variance was used to assess vegetation-type and seasonal effects. Principal component analysis (PCA) was used to determine the similarities between succession and native systems according to fatty acid concentration (mole %). Pearson's correlation coefficients were used to assess correlations between microbial parameters and soil properties. Statistical analyses were considered significant at $p < 0.05$. Statistical analyses were performed using SPSS v12.0 (SPSS Inc., Chicago, IL, USA).

RESULTS

Chemical and Physical Properties of Soils

Soil moisture was higher in the summer than in the winter (Table 2) because winters are dry in this area. The CONIF-40 forest contained the highest soil moisture content in the summer. The soil organic C and total N contents were highest in the CONIF-40 forest, with no significant differences between the CONIF-80 and BROAD-Nat forests. All soils were strongly acidic.

Table 2: Soil chemical and physical properties and microbial biomass in 40-year-old (CONIF-40) and 80-year-old (CONIF-80) coniferous plantations and a natural broadleaf forest (BROAD-Nat) in two seasons

Season	Forest Type	Soil Moisture (%)	pH	C_{org} [1] (%)	N_{tot} (%)	C_{mic} ($mg \cdot kg^{-1}$)	N_{mic} ($mg \cdot kg^{-1}$)	C_{mic}/C_{org} (%)	N_{mic}/N_{tot} (%)
Summer	BROAD-Nat	27.6 a [2]	3.78 b	3.45 b	0.29 ab	1668 a	140 a	4.83 a	4.82 b
	CONIF-80	23.1 b	3.74 b	3.23 bc	0.24 bc	1314 b	94.5 c	4.21 a	3.93 c
	CONIF-40	29.2 a	3.62 c	4.45 a	0.31 a	1229 b	116 b	2.82 b	3.74 c
Winter	BROAD-Nat	19.3 c	3.87 a	2.62 c	0.22 c	763 c	119 b	2.92 b	5.40 a
	CONIF-80	21.9 c	3.87 a	3.31 bc	0.25 bc	748 c	121 ab	2.28 c	4.84 b
	CONIF-40	27.3 a	3.75 b	3.79 b	0.29 ab	812 c	117 b	2.16 c	3.98 c
ANOVA results [3]									
Forest type		***	***	**	**	***	***	***	***
Sample season		***	***	ns	*	***	ns	***	***
Forest × season		***	ns	ns	*	***	***	ns	ns

[1] C_{org}: soil organic C, N_{tot}: soil total N, C_{mic}: microbial biomass C, N_{mic}: microbial biomass N; [2] Values in each column followed by different lowercase letters are significantly different at $p < 0.05$, based on Duncan's multiple test; [3] * $p < 0.05$, ** $p < 0.01$, *** $p < 0.001$, ns: not significant.

Soil Microbial Biomass

Soil C_{mic} and N_{mic} contents were significantly higher in the summer than in the winter. Soil C_{mic} and N_{mic} contents did not differ significantly among the three forests in the winter. During the summer, both soil C_{mic} and N_{mic} contents were higher in the BROAD-Nat forest than in the CONIF-40 and CONIF-80 forests. Ratios of C_{mic}/C_{org} and N_{mic}/N_{tot} were significantly different among the different forest soils, being higher in the BROAD-Nat forest than in the CONIF-40 and CONIF-80 forests (Table 2).

Soil Enzyme Activities

Soil xylanase, phosphatase, and urease activities were significantly affected by forest type and season. Summer soil cellulase, xylanase, and phosphatase activities were significantly higher in the CONIF-40 forest than in the BROAD-Nat forest (Table 3), with no significant difference between the CONIF-80 and BROAD-Nat forests. Soil enzyme activities in the winter did not differ among forest types, except for a lower xylanase activity in the CONIF-80 forest. Additionally, enzymatic activities were higher in the summer than in winter, except for xylanase.

Table 3: Soil enzymatic activities in 40-year-old (CONIF-40) and 80-year-old (CONIF-80) coniferous forest plantations and a natural broadleaf forest (BROAD-Nat) in two seasons

Season	**Forest Type**	**Cellulase (μg glucose $g^{-1}\cdot h^{-1}$)**	**Xylanase (μg glucose $g^{-1}\cdot h^{-1}$)**	**Urease (mmol NH_4^+-N $g^{-1}\cdot h^{-1}$)**	**Acid Phosphatase (μg nitrophenol $g^{-1}\cdot h^{-1}$)**
Summer	BROAD-Nat	31.7 c [1]	65.8 d	10.6 a	569 b
	CONIF-80	63.8 bc	70.8 cd	11.9 a	628 b
	CONIF-40	126 a	109 ab	9.47 ab	850 a
Winter	BROAD-Nat	36.7 bc	116 a	5.83 b	338 c
	CONIF-80	46.7 bc	92.1 bc	9.84 ab	318 c
	CONIF-40	76.2 b	118 a	6.45 b	361 c
ANOVA results					
Forest type		**	*	*	ns
Sample Season		ns	***	**	***
Forest × season		ns	ns	Ns	ns

[1] Values in each column followed by different lowercase letters are significantly different at $p < 0.05$, based on Duncan's multiple test; * $p < 0.05$, ** $p < 0.01$, *** $p < 0.001$, ns: not significant.

Microbial Community Structure

Phospholipid acid biomarkers were significantly affected by forest type and season. In the summer, the total soil PLFA content as well as contents of bacteria, VAM fungi, actinomycetes, and G+ and G− bacteria were significantly lower in CONIF-40 and CONIF-80 samples than in

BROAD-Nat samples (Table 4). Ratios of fungi to bacteria and of G+ to G− bacteria were also higher in CONIF-40 forest than in BROAD-Nat forest. Contents of total soil PLFA, bacteria, fungi, VAM fungi, actinomycetes, G+ bacteria, and G− bacteria were all lower in the winter than in the summer for all forest types, and values did not significantly differ among the three forests in the winter. However, as in summer, the ratio of fungi to bacteria was higher in the CONIF-40 forest than in the BROAD-Nat forest in the winter, while this ratio was similar between the CONIF-80 and BROAD-Nat forests during both seasons. The ratio of soil G+ to G− bacteria was higher in the winter than in the summer.

The PCA based on PLFA soil microbial communities in the summer indicated that the first principal component (PC1) and second principal component (PC2) accounted for 62% of the forest soil variation. The high loadings in PC1 included G+ bacteria (i15:0, a15:0, i16:0, and 10Me16:0; positive loadings) and G− bacteria (18:1ω7c and cy19:0; negative loadings) (Figure 2). The high loading in PC2 was a result of the actinomycetes (10Me17:0; positive loading). The majority of the variability was due to the differences between the CONIF-40 forest samples and those of CONIF-80 and BROAD-Nat forest. In this study, PC1 explains 43% of variability, and separates the CONIF-40 samples (positive loadings) from the CONIF-80 and BROAD-Nat samples (negative loadings). The PC2, which explains 19% of the variability, separates the samples CONIF-80 (negative loadings) from BROAD-Nat (positive loadings) (Figure 3). The CONIF-80 forest differed from the BROAD-Nat forest; however, this difference was smaller than that between CONIF-80 and CONIF-40. In winter, differences in the PLFA profiles of the three forest types were not as clear (Figure 4).

Table 4: Contents of phospholipid acid biomarkers (nmol·g^{-1} soil) and biomarker content ratios in 40-year-old (CONIF-40) and 80-year-old (CONIF-80) coniferous forest plantations and a natural broadleaf forest (BROAD-Nat) in two seasons

Season	Forest Type	Total PLFA [1]	Bacteria	Fungi	VAM Fungi	Actino-mycetes	G+	G–	G+/G–	Fungi/Bacteria
Summer	BROAD-Nat	115 a [2]	50.4 a	1.92 a	2.59 a	3.53 a	23.4 a	34.1 a	0.69 c	0.038 b
	CONIF-80	80.7 b	34.5 b	1.63 a	1.68 b	1.99 b	16.6 b	23.4 b	0.71 c	0.047 ab
	CONIF-40	71.5 bc	29.8 b	1.59 a	1.38 b	1.80 b	17.0 b	16.4 c	1.16 b	0.052 a
Winter	BROAD-Nat	50.7 d	12.5 c	0.270 b	0.611 c	0.781 c	6.92 d	5.24 d	1.32 ab	0.021 c
	CONIF-80	58.7 d	17.4 c	0.420 b	0.650 c	0.790 c	9.80 c	7.14 d	1.37 ab	0.024 c
	CONIF-40	64.1 d	18.4 c	0.701 b	0.705 c	0.777 c	10.7 c	7.02 d	1.53 a	0.038 b
ANOVA results [3]										
Forest type		*	*	ns	**	***	*	**	***	***
Sampling season		***	***	***	***	***	***	***	***	***
Forest × season		**	**	ns	***	***	***	***	ns	ns

[1] PLFA: phospholipid fatty acids, VAM: vesicular-arbuscular mycorrhizal, G+: gram-positive bacteria, G–: gram-negative bacteria; [2] Values in each column followed by different lowercase letters are significantly different at $p < 0.05$, based on Duncan's multiple test; [3] * $p < 0.05$, ** $p < 0.01$, *** $p < 0.001$, ns: not significant.

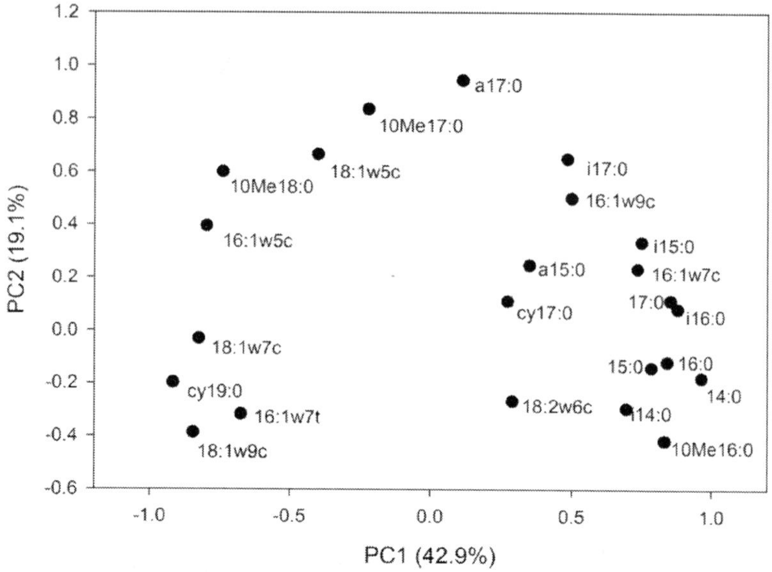

Figure 2: Corresponding loading values from the principal component (PC) analysis of the mole % microbial phospholipid fatty acid content in soil samples from three forest types in summer.

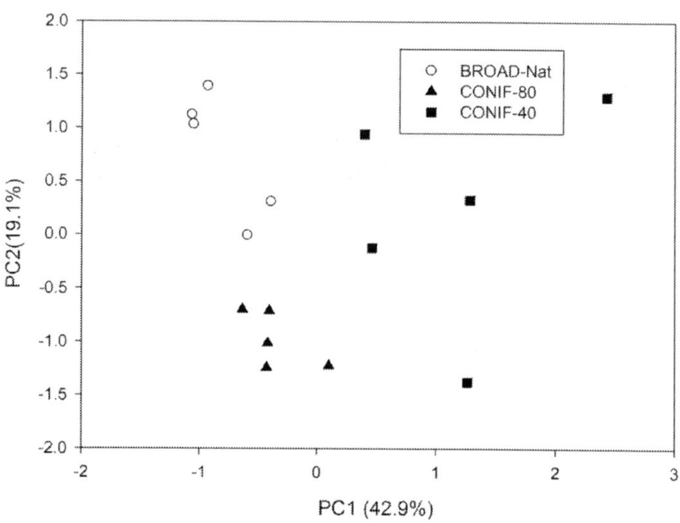

Figure 3: Plots of the two first principal components (PCs) from the principal component analysis of the mole % microbial phospholipid fatty acid content in soil samples from three forest types in summer.

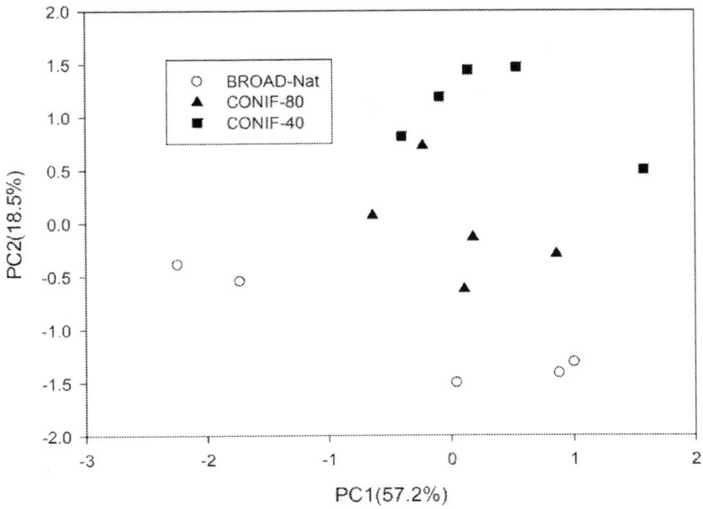

Figure 4: Plots of the two first principal components (PCs) from the principal component analysis of the mole % microbial phospholipid fatty acid content in soil samples from three forest types in winter.

Correlations

Soil cellulase and phosphatase activities were positively correlated with C_{org} content ($p < 0.05$; Table 5). The total PLFA and G+ bacterial contents as well as soil enzyme activities, except xylanase, were positively correlated with soil N_{tot} content. Additionally, total PLFA content was positively correlated with C_{mic} content, while xylanase activity was negatively correlated with C_{mic} content. PLFA marker content, excluding fungal markers, was positively correlated with the C_{mic}/C_{org} ratio and soil moisture.

Table 5: Pearson's correlation coefficients between soil properties and microbial variables

Soil Properties	Cellulase	Xylanase	Urease	Phosphatase	Total PLFA	G+	G−	G+/G−	Fungi	VAM Fungi	Fungi/ Bacteria
C	0.643 ***	0.292	0.366	0.454 *	0.320	0.376	0.177	−0.050	0.132	0.186	0.159
N	0.476 **	0.118	0.406 *	0.394 *	0.464 *	0.498 **	0.325	−0.050	0.132	0.339	0.206
C_{mic}	−0.039	−0.606 ***	0.503 **	0.619 ***	0.895 ***	0.917 ***	0.964 ***	−0.902 ***	−0.651 ***	0.952 ***	0.888 ***
C_{mic}/C_{org}	−0.365	−0.753 **	0.268	0.363	0.711 ***	0.711 ***	0.843 ***	−0.884 ***	−0.672 ***	0.821 ***	0.776 ***
C_{mic}/N_{mic}	0.112	−0.584 **	0.459 *	0.676 ***	0.665 ***	0.743 ***	0.819 ***	−0.889 ***	0.670 ***	0.779 ***	0.919 ***
pH	−0.678 ***	−0.148	−0.273	−0.682 ***	−0.419 *	−0.513 **	−0.381 *	0.247	0.202	−0.389 *	−0.492 **
Moisture	0.428 *	−0.039	0.291	0.529 **	0.637 ***	0.680 **	0.529 **	−0.252	−0.023	0.556 **	0.440 *

* $p < 0.05$, ** $p < 0.01$, *** $p < 0.001$.

DISCUSSION

Previous studies involving soil organic matter characterization using ^{13}C nuclear magnetic resonance (NMR) at the same study site showed greater recalcitrant tannin and polyphenol contents in coniferous plantations (*C. konishii* and *C. formosana*) than in BROAD-Nat forest [28]. The resistance of coniferous litter to decomposition may contribute to the accumulation of C and N in soil. However, a decrease in the quality of organic matter input after the replacement of the native broadleaf vegetation with a pure coniferous plantation led to reductions in microbial C and N biomasses (Table 2). Swift et al. [29] showed that litter with a high C/N ratio, such as in coniferous forests, was a poor substrate for microbial growth. Both the C_{mic} and N_{mic} contents were high in the summer, low in the winter, and positively correlated with soil moisture (Table 5). Cool temperatures combined with low rainfall were responsible for the low microbial biomass in winter. Liu et al. [30] reported that the lack of water and low temperatures in winter reduced the microbial biomass in a tropical forest in China.

Several studies have shown that enzyme activities depend on the quality of organic matter [31,32]. Thus, differences in enzyme activities among forest types may have been due to differences in organic matter quality. The leaf litter in native broadleaf forests contains fewer components that are resistant to decomposition, leading to a faster decomposition rate than that of conifer litter [33]. The cellulase activity was positively correlated with the soil organic matter content and negatively correlated with pH (Table 5), in agreement with the results of Safari and Sharifi [34].

Community-level PLFA profiles are useful for detecting soil microbial community responses to various land uses or disturbances in ecosystems [35,36]. The litter derived from coniferous forests decomposes slowly; therefore, coniferous forests may provide fewer labile compounds than the native broadleaf forest. The G− bacterial content has been reported to increase with higher substrate availability [37,38]. Additionally, Kelly et al. [39] reported faster growth of G− compared to that of G+ bacteria in fertile environments. Not surprisingly, the G− bacterial content in our study was lower in coniferous soil than in broadleaf soil. G+ bacteria may be able to adapt to inferior conditions more effectively than G− bacteria, resulting in higher G+/G− ratios. The clearing of native forests and the conversion to coniferous species leads to environmental stress, which may explain the higher G+/G− ratio in CONIF-40 soil compared with that in BROAD-Nat soil. Likewise, Ushio et al. [7,40] reported a higher G+/G− ratio in coniferous soils than in hardwood soils. Similarities in soil G+/G− ratios between CONIF-80 and BROAD-Nat forests, however, confirm our hypothesis that the microbial community structure can be restored with long-term succession despite the use of a different vegetation type through the development of an understory. Ample litter from broadleaf tree regrowth in the CONIF-80 forest may have reduced differences in litter quality between CONIF-80 and BROAD-Nat forests. Seasonally, the higher G+/G− ratio in the winter than in the summer is also due to the greater adaption of G+ bacteria to unfavorable conditions. Generally, G− bacteria are more sensitive to changes in soil moisture [41]. Schimel et al. [42] indicated that drought-tolerant microbes, such as G+ bacteria, are inherently protected from low moisture during drought stress.

The ratio of fungi to bacteria increased in coniferous forest soils after the conversion from BROAD-Nat forest (Table 4). Several studies have shown that the microbial community structure responds to soil fertility, with fertile soils favoring bacteria while less fertile soils favor fungi [43,44]. High availability of labile C, for which bacteria are more competitive than fungi, may restrain fungal growth in natural broadleaf forest soil [33]. Greater concentrations of tannins in coniferous forests may also contribute to a higher ratio of fungi to bacteria, as fungi have a higher tannin tolerance [40]. Additionally, fungi can use recalcitrant compounds as C sources [45]. Ushio et al. [7] found the ratio of fungi to bacteria to be significantly higher in coniferous compared to broadleaf soils in tropical mountain areas. In this study, the higher soil fungal content in CONIF-40 than in the BROAD-Nat forest and the similarities in the soil fungal contents between CONIF-80 and the BROAD-Nat forest suggest an evolution of the microbial community that coincides with forest clearing, reforestation, and aging of secondary forest. Thus, forest clearing and reforestation with coniferous species changes the microbial community by increasing soil fungi and decreasing soil bacteria, resulting in a high ratio of fungi to bacteria in the soil. The ratio of soil fungi to bacteria, however, returns to levels similar to that of the native forest system with the long-term (~80 years) growth of secondary forest. Understory broadleaf shrubs and trees that gradually become established over time eventually contribute sufficient litter to balance out the effects of the coniferous litter, favoring the recovery of bacteria.

The 16:1ω5 PLFA peak is an indicator of VAM fungi [46]. The highest soil VAM fungal content was observed in the BROAD-Nat forest in the summer, suggesting that broadleaf plants support mycorrhizal fungi [47]. Priha et al. [48] and Djukic et al. [49] also found higher contents of 16:1ω5 PLFA peak in broadleaf- than in conifer-dominated forest soils. However, one should be careful in using 16:1ω5 PLFA peak as the marker for AM fungi. Olsson [50] indicated this PLFA peak could be also found in some bacteria. Priha et al. [51] noted the 16:1ω5 PLFA peak content may respond to changes in the soil available C. Thus, the high 16:1ω5 content in the BROAD-Nat forest may be due to the high labile soil C in the broadleaf forest.

The PCA of the PLFA content clearly differentiated between the CONIF-40 forest soil and the other two forest soils, largely due to the PC1 axis. The bacterial biomarkers contributed the most to the PC1 score, suggesting that bacterial content is an important factor in determining the microbial community structure of forests in this ecosystem. Our results indicate that the soil microbial community under the CONIF-40 forest was still in a successional phase, while the soil microbial community under the CONIF-80 forest more reached that of the BROAD-Nat forest.

CONCLUSIONS

Soil microbial biomass and enzyme activities remain lower in secondary coniferous forests than in a natural broadleaf forest even after several decades of vegetation replacement. However, 80 years of growth of a secondary conifer forest led to a soil microbial community structure similar to that under the natural broadleaf forest. We found that the soil microbial community changes along with the growth of coniferous forest in a subtropical region. The conversion of broadleaf to coniferous forests leads to increased ratios of fungi to bacteria and of G+ to G− bacteria. Following the long-term aging of the coniferous plantation, however, bacterial and G− bacterial contents increased, returning the soil microbial community to lower ratios of fungi to bacteria and of G+ to G− bacteria and restoring values similar to those in the native broadleaf forest. The understory broadleaf shrubs and trees established after the reforestation of coniferous plantations may balance out the effects of the coniferous litter, contributing to the recovery of bacteria. Such a mechanism may improve soil restoration after reforestation in subtropical areas.

ACKNOWLEDGMENTS

This work was funded by the Ministry of Science and Technology, Taiwan (MOST 104-2621-B-001-005) and by Academia Sinica.

AUTHOR CONTRIBUTIONS

E.-H.C., T.-H.C., and C.-Y.C. conceived and designed the experiments; E.-H.C. performed the experiments; E.-H.C. and G.T. analyzed the data; C.-K.H. contributed plant inventory; E.-H.C. wrote the original draft; G.T. and C.-Y.C. reviewed and edited the manuscript.

REFERENCES

1. Iovieno, P.; Alfani, A.; Baath, E. Soil microbial community structure and biomass as affected by *Pinus pinea*plantation in two Mediterranean areas. *Appl. Soil Ecol.* 2010, *45*, 56–63.
2. Chodak, M.; Niklińska, M. The effect of different tree species on the chemical and microbial properties of the reclaimed mine soils. *Biol. Fertil. Soils* 2010, *46*, 555–566.
3. Samonil, P.; Král, K.; Hort, L. The role of tree uprooting in soil formation: A critical literature review. *Geoderma*2010, *157*, 65–79.
4. Banning, N.C.; Gleeson, D.B.; Grigg, A.H.; Grant, C.D.; Andersen, G.L.; Brodie, E.L.; Murphy, D.V. Soil microbial community successional patterns during forest ecosystem restoration. *Appl. Environ. Microbiol.* 2011, *7*, 6158–6164.
5. Zheng, H.; Ouyang, Z.Y.; Wang, X.K.; Fang, Z.G.; Zhao, T.Q.; Miao, H. Effects of regenerating forest cover on soil microbial communities: A case study in hilly red soil region, Southern China.

For. Ecol. Manag. 2005, *217*, 244–254.

6. Cao, Y.S.; Fu, S.L.; Zou, X.M.; Cao, H.L.; Shao, Y.H.; Zhou, L.X. Soil microbial community composition under Eucalyptus plantations of different age in subtropical China. *Eur. J. Soil Biol.* 2010, *46*, 128–135.
7. Ushio, M.; Wagai, R.; Balser, T.C.; Kitayama, K. Variations in the soil microbial community composition of tropical montane forest ecosystem: Does tree species matter? *Soil Biol. Biochem.* 2008, *40*, 2699–2702.
8. Moore-Kucera, J.; Dick, R.P. PLFA profiling of microbial community structure and seasonal shifts in soils of a Douglas-fir chronosequence. *Microb. Ecol.* 2008, *55*, 500–511.
9. Lou, X.; Cui, X.Y. The study of soil microbial diversity in temperate forest secondary succession process. *Chin. J. Soil Sci.* 2011, *3*, 010.
10. Overby, S.T.; Hart, S.C. Short-term belowground responses to thinning and burning treatments in southwestern Ponderosa Pine Forest of the USA. *Forest* 2016, *7*, 45.
11. Chan, O.C.; Yang, X.; Fu, Y.; Feng, Z.; Sha, L.; Casper, P.; Zou, X. 16S rRNA gene analyses of bacterial community structures in the soils of evergreen broad-leaved forests in south-west China. *FEMS Microbiol. Ecol.* 2006, *58*, 247–259.
12. Lin, Y.T.; Whitman, W.B.; Coleman, D.C.; Chiu, C.Y. Molecular characterization of soil bacterial community in a perhumid, low mountain forest. *Microbe. Environ.* 2011, *26*, 325–331.
13. Weand, M.P.; Arthur, M.A.; Lovett, G.M.; McCulley, R.L.; Weathers, K.C. Effects of tree species and N additions on forest floor microbial communities and extracellular enzyme activities. *Soil Biol. Biochem.* 2010, *42*, 2161–2173.
14. Wang, X.; Wang, X.; Zhang, W.; Shao, Y.; Zou, X.; Liu, T.; Zhou, L.; Wan, S.; Rao, X.; Li, Z.; et al. Invariant community structure of soil bacteria in subtropical coniferous and broadleaved forests. *Sci. Rep.* 2016, *6*, 19071.
15. Lin, Y.T.; Jangid, K.; Whitman, W.B.; Coleman, D.C.; Chiu, C.Y. Change in bacterial community structure in response to disturbance of natural hardwood and secondary coniferous forest soils in central Taiwan. *Microb. Ecol.* 2011, *61*, 429–437.

16. Zelles, L. Fatty acid patterns of phospholipids and lipopolysaccharides in the characterization of microbial communities in soil: A review. *Biol. Fertil. Soils* 1999, *29*, 111–129.

17. Farrell, M.; Griffith, G.W.; Hobbs, P.J.; Perkins, W.T.; Jones, D.L. Microbial diversity and activity are increased by compost amendment of metal-contaminated soil. *FEMS Microbiol. Ecol.* 2010, *71*, 94–105.

18. McLean, E.O. Soil pH and lime requirement. In *Methods of Soil Analysis, Part 2 Chemical and Microbiological Properties*; Page, A.L., Ed.; American society of Agronomy: Madison, WI, USA, 1982; pp. 199–224.

19. Vance, E.D.; Brookes, P.C.; Jenkinson, D.S. An extraction method for measuring soil microbial biomass C. *Soil Biol. Biochem.* 1987, *19*, 703–707.

20. Wu, J.; Joergensen, R.G.; Pommerening, B.; Chaussod, R.; Brookes, P.C. Measurement of soil microbial biomass C by fumigation-extraction-an automated procedure. *Soil Biol. Biochem.* 1990, *22*, 1167–1169.

21. Amato, M.; Ladd, J.N. Assay for microbial biomass based on ninhydrin-reactive nitrogen in extracts of fumigated soils. *Soil Biol. Biochem.* 1988, *20*, 107–114.

22. Kandeler, E.; Gerber, H. Short-term assay of soil urease activity using colorimetric determination of ammonium. *Biol. Fertil. Soils* 1988, *8*, 199–202.

23. Tabatabai, M.A.; Bremner, J.M. Use of p-nitrophenyl phosphate for assay of soil phosphatase activity. *Soil Biol. Biochem.* 1969, *1*, 301–307.

24. Schinner, F.; Von Mersi, W. Xylanase-, CM-cellulase- and invertase activity in soil: An improved method. *Soil Biol. Biochem.* 1990, *22*, 511–515.

25. Frostegard, A.; Tunlid, A.; Baath, E. Phospholipid fatty acid composition, biomass, and activity of microbial communities from two soil types exposed to different heavy metals. *Appl. Environ. Microbiol.* 1993, *59*, 3605–3617.

26. Chang, E.H.; Chen, C.T.; Chen, T.H.; Chiu, C.Y. Soil microbial communities and activities in sand dunes of subtropical coastal

forests. *Appl. Soil Ecol.* 2011, *49*, 256–262.

27. Zogg, G.P.; Zak, D.R.; Ringleberg, D.B.; MacDonald, N.W.; Pregitzer, K.S.; White, D.C. Compositional and functional shifts in microbial communities due to soil warming. *Soil Sci. Soc. Am. J.* 1997, *61*, 475–481.
28. Jien, S.H.; Chen, T.H.; Chiu, C.Y. Effects of afforestation on soil organic matter characteristics under subtropical forests with low elevation. *J. For. Res.* 2011, *16*, 275–283.
29. Swift, M.J.; Heal, O.W.; Anderson, J.M. *Decomposition in Terrestrial Ecosystems*; Blackwell Scientific: Oxford, UK, 1979.
30. Liu, L.; Gundersen, P.; Zhang, T.; Mo, J.M. Effects of phosphorus addition on soil microbial biomass and community composition in three forest types in tropical China. *Soil Biol. Biochem.* 2012, *44*, 31–38.
31. Nannipieri, P.; Muccini, L.; Ciardi, C. Microbial biomass and enzyme activities: Production and persistence. *Soil Biol. Biochem.* 1983, *15*, 679–685.
32. Badiane, N.N.Y.; Chotte, J.L.; Pate, E.; Masse, D.; Rouland, C. Use of soil enzyme activities to monitor soil quality in natural and improved fallows in semi-arid tropical regions. *Appl. Soil Ecol.* 2001, *18*, 229–238.
33. Yang, Y.S.; Guo, J.F.; Chen, G.S.; Xie, J.S.; Cai, L.P.; Lin, P. Litterfall, nutrient return, and leaf-litter decomposition in four plantations compared with a natural forest in subtropical China. *Ann. For. Sci.* 2004, *61*, 465–476.
34. Safari, S.A.A.; Sharifi, Z. Factors affecting cellulase activity in some soils in Hamadan Province, Iran. *Iran J. Agric. Sci.* 2006, *37*, 645–652.
35. MacKenzie, M.D.; Quideau, S.A. Microbial community structure and nutrient availability in oil sands reclaimed boreal soils. *Appl. Soil Ecol.* 2010, *44*, 32–41.
36. Kytöviita, M.M.; Pietikäinen, A.; Fritze, H. Soil microbial and plant responses to the absence of plant cover and monoculturing in low arctic meadows. *Appl. Soil Ecol.* 2011, *48*, 142–151.
37. Zelles, L.; Bai, Q.Y.; Beck, T.; Beese, F. Signature fatty acids in

phospholipids and lipopolysaccharides as indicators of microbial biomass and community structure in agricultural soils. *Soil Biol. Biochem.* 1992, *24*, 317–332.

38. Bossio, D.A.; Scow, K.M.; Gunapala, N.; Graham, K.J. Determinants of soil microbial communities: Effects of agricultural management, season, and soil type on phospholipid fatty acid profiles. *Microb. Ecol.* 1998, *36*, 1–12.
39. Kelly, J.J.; Favila, E.; Hundal, L.S.; Marlin, J.C. Assessment of soil microbial communities in surface applied mixtures of Illinois River sediments and biosolids. *Appl. Soil Ecol.* 2007, *36*, 176–183.
40. Ushio, M.; Kitayama, K.; Balser, T.C. Tree species-mediated spatial patchiness of the composition of microbial community and physicochemical properties in the topsoils of a tropical montane forest. *Soil Biol. Biochem.* 2010, *42*, 1588–1595.
41. Wilkinson, S.C.; Anderson, J.M.; Scardelis, S.P.; Tisiafouli, M.; Taylor, A.; Wolters, V. PLFA profiles of microbial communities in decomposing conifer litters subject to moisture stress. *Soil Biol. Biochem.* 2002, *34*, 189–200.
42. Schimel, J.P.; Balser, T.C.; Wallenstein, M. Microbial stress-response physiology and its implications for ecosystem function. *Ecology* 2007, *86*, 1386–1394.
43. Grayston, S.J.; Campbell, C.D.; Bardgett, R.D.; Mawdsley, J.L.; Clegg, C.D.; Ritz, K.; Griffiths, B.S.; Rodwell, J.S.; Edwards, S.J.; Davies, W.J.; et al. Assessing shifts in microbial community structure across a range of grasslands of differing management intensity using CLPP, PLFA and community DNA techniques. *Appl. Soil Ecol.* 2004, *25*, 63–84.
44. Grayston, S.J.; Prescott, C.E. Microbial communities in forest floors under four tree species in coastal British Columbia. *Soil Biol. Biochem.* 2005, *37*, 1157–1167.
45. Scalbert, A. Antimicrobial properties of tannins. *Phytochemistry* 1991, *30*, 3875–3883.
46. Olsson, P.A.; Baath, E.; Jakobsen, I.; Soderstrom, B. The use of phospholipid and neutral lipid fatty acids to estimate biomass of arbuscular mycorrhizal fungi in soil. *Mycol. Res.* 1995, *99*, 623–629.
47. Kourtev, P.S.; Ehrenfeld, J.G.; Haggblom, M. Experimental analysis

of the effect of exotic and native plant species on the structure and function of soil microbial communities. *Soil Biol. Biochem.* 2003, *35*, 895–905.

48. Priha, O.; Grayston, S.J.; Hiukka, R.; Pennanen, T.; Smolander, A. Microbial community structure and characteristics of the organic matter in soils under *Pinus sylvestris*, *Picea abies*, and *Betula pendula* at two forest sites. *Biol. Fertil. Soils* 2001, *33*, 17–24.
49. Djukic, I.; Zehetner, F.; Mentler, A.; Gerzabek, M.H. Microbial community composition and activity in different Alpine vegetation zones. *Soil Biol. Biochem.* 2010, *42*, 155–161.
50. Olsson, P.A. Signature fatty acids provide tools for determination of the distribution and interactions of mycorrhizal fungi in soil. *FEMS Microbiol. Ecol.* 1999, *29*, 303–310.
51. Priha, O.; Grayston, S.J.; Pennanen, T.; Smolander, A. Microbial activities related to C and N cycling and microbial community structure in the rhizospheres of *Pinus sylvestris*, *Picea abies* and *Betula pendula* seedlings in an organic and mineral soil. *FEMS Microbiol. Ecol.* 1999, *30*, 187–199.

CITATIONS

CHAPTER 7

Schloss PD, Handelsman J (2006) Toward a Census of Bacteria in Soil. PLoS Comput Biol 2(7): e92. https://doi.org/10.1371/journal.pcbi.0020092.

CHAPTER 8

Lucas William Mendes and Siu Mui Tsai, Variations of Bacterial Community Structure and Composition in Mangrove Sediment at Different Depths in Southeastern Brazil, doi:10.3390/d6040827.

CHAPTER 9

Stefani FOP, Bell TH, Marchand C, de la Providencia IE, El Yassimi A, St-Arnaud M, et al. (2015) Culture-Dependent and -Independent Methods Capture Different Microbial Community Fractions in Hydrocarbon-Contaminated Soils. PLoS ONE 10(6): e0128272. https://doi.org/10.1371/journal.pone.0128272.

CHAPTER 10

Fisseha Asmelash, Tamrat Bekele, and Emiru Birhane, The Potential Role of Arbuscular Mycorrhizal Fungi in the Restoration of Degraded Lands, doi: 10.3389/fmicb.2016.01095.

CHAPTER 11

Ed-Haun Chang, Tsai-Huei Chen, Guang-long Tian, Chun-Kai Hsu and Chih-Yu Chiu, Effect of 40 and 80 Years of Conifer Regrowth on Soil Microbial Activities and Community Structure in Subtropical Low Mountain Forests, doi:10.3390/f7100244.

INDEX